Politics of Last Resort

Politics of Last Resort

Governing by Emergency in the European Union

JONATHAN WHITE

OXFORD
UNIVERSITY PRESS

OXFORD
UNIVERSITY PRESS

Great Clarendon Street, Oxford, OX2 6DP,
United Kingdom

Oxford University Press is a department of the University of Oxford.
It furthers the University's objective of excellence in research, scholarship,
and education by publishing worldwide. Oxford is a registered trade mark of
Oxford University Press in the UK and in certain other countries

First Edition published in 2020

Impression: 1

Published in the United States of America by Oxford University Press
198 Madison Avenue, New York, NY 10016, United States of America

British Library Cataloguing in Publication Data
Data available

Library of Congress Control Number: 2019947622

ISBN 978–0–19–879172–0

DOI: 10.1093/oso/9780198791720.001.0001

Printed and bound in Great Britain by
Clays Ltd, Elcograf S.p.A.

For Hana

Acknowledgements

The book has benefitted immensely from the input of others. Some read large chunks in draft form: I am deeply grateful to Carlo Invernizzi Accetti, Waltraud Schelkle, Jan Suntrup, and Lea Ypi for their willingness to devote time and offer valuable suggestions. Others encountered the argument at workshops and seminars, and for their comments I would particularly like to thank the following: Bruce Ackerman, Štefan Auer, Marija Bartl, Olivier Beaud, Richard Bellamy, Vojtěch Belling, Chris Bickerton, Jacco Bomhoff, Hauke Brunkhorst, Dario Castiglione, Damian Chalmers, Ben Crum, Deirdre Curtin, Marco Dani, Mark Dawson, Pieter de Wilde, Dimitris Efthymiou, Erik Oddvar Eriksen, John Erik Fossum, Simon Glendinning, Sverker Gustavsson, Lise Herman, Josef Hien, Sara Hobolt, Jonathan Hopkin, Achim Hurrelmann, Daniel Innerarity, Abby Innes, Giulio Itzcovich, Christian Joerges, Andreas Kalyvas, Lucy Kinski, Mareike Kleine, Satoshi Kodama, Jan Komárek, Christian Kreuder-Sonnen, Joseph Lacey, Christopher Lord, Philip Manow, Agustín Menéndez, Benjamin Moffitt, Glyn Morgan, Jan-Werner Müller, Kalypso Nicolaidis, Pieter Niesen, Claus Offe, Markus Patberg, Anne Phillips, Sonja Puntscher Riekmann, Mark Salter, Bill Scheuerman, Nicole Scicluna, Astrid Séville, Alexander Somek, Antoine Vauchez, Albert Weale, Frank Wendler, Mike Wilkinson, Fabio Wolkenstein, and David Woodruff. I also benefitted greatly from two outstanding PhD researchers working on related topics in this period: Hjalte Lokdam and Marta Lorimer.

The idea for the book emerged in 2012, when writing what would later be published as 'Emergency Europe' in *Political Studies*. The article was given the journal's 2015 Harrison Prize for best paper, and I am very grateful to the editors and jury for the encouragement to go further into the topic. Most of the writing for the book was done based at my usual academic home, the LSE's European Institute. I thank staff and students for making it a friendly and stimulating environment. I would also like to acknowledge the Wissenschafts-kolleg zu Berlin, where ideas were developed during a one-year fellowship in 2014–15 funded by the EURIAS Foundation, and the Europe Center at Stanford University, where a chapter was written during a fellowship in spring 2017.

At Oxford University Press I would like to thank Dominic Byatt, as well as three anonymous reviewers. Two chapters draw on previously published material, and I acknowledge Wiley for permission to reproduce passages from the following: 'Principled Disobedience in the EU', *Constellations* 24 (4), 2017; and 'Authority after Emergency Rule', *Modern Law Review* 78 (4), 2015.

Family life with three young children is a good way to get thinking about permanent states of emergency. Sharing an increasingly noisy house with Arbien, Rubin, and Hana has been a source of pleasure and inspiration. I am very grateful to them in countless ways, as I am to my mother Ann and my late father John. In particular I would like to thank Lea, for her conversation, love, sacrifices, and support.

Contents

Introduction 1

1. Beyond a Sovereign State of Exception 16
2. Structural Vulnerabilities of the Transnational Sphere 40
3. Towards Emergency Europe 64
4. Prospects of a New Normal 86
5. Iterations of Technocracy 106
6. Populism as the Promise of Agency 127
7. Principled Disobedience 147
8. Structuring Partisan Power 167
9. Beyond a Politics of Singularities 188

Bibliography 205
Index 225

Introduction

> 'The challenges of recent years forced national governments and EU institutions to take quick and extraordinary steps. They needed to stabilise their economies and to protect all that has been achieved through the gradual and at times painstaking process of European integration. As a result, the integrity of the Euro area as a whole has been preserved and the internal market remains strong.... Europe's Economic and Monetary Union (EMU) today is like a house that was built over decades but only partially finished. When the storm hit, its walls and roof had to be stabilised quickly. It is now high time to reinforce its foundations...'
>
> The 'Five Presidents' Report',
> European Commission (2015), p. 4.

When the art of governing is cast as the taming of extreme natural forces, decision-making predictably resembles the management of an emergency. From storms in the economy to waves of migration, themes of freakish weather are part of the vernacular of modern politics. As some officials battle the elements, others contend with gravity or infection: markets risk collapse, currencies, and confidence plummet, and contagion breaks out. Authorities claim the need to act fast and decisively. 'Quick and extraordinary steps' aim to stabilize the situation, containing what threatens to become uncontainable. Then come the preparations for things yet to happen: foundations are reinforced, defences mended, shock absorbers installed. With some intensity in the 2010s, Europe's executive powers came to embrace this way of governing, with actions departing from convention rationalized as necessary responses to exceptional and urgent threats.

Looking back on developments in Europe over the last decade and more, frenetic activity and the improvisation of new arrangements are some of the dominant themes. The European Union (EU) has undergone great change. Interventions by national and European executives to control money and people have left a legacy of new institutions that continues to expand. From economic monitoring to border policing, one sees new formations spanning the national and supranational that reconfigure how decisions are made and enforced. Introduced and maintained in the name of crisis response, these structures well exceed what was envisioned in prior formalizations of the EU system, and challenge the norms of national politics too.

Politics of Last Resort: Governing by Emergency in the European Union. Jonathan White, Oxford University Press (2020).
DOI: 10.1093/oso/9780198791720.001.0001

And yet, while the institutional scene has been in persistent flux, the substance of policy and the contours of social power look more constant. As many accounts on the theme of the 'strange non-death of neoliberalism' suggest, the basics of the economic order have been slow to change. Principles of low deficit, low spending, and competition continue to be embraced as the foundation for public policy, while the prioritization of debt repayment continues to be axiomatic. Border regimes continue to be geared to an idea of fortress Europe, now pursued more stringently. In some ways the capacity to contest these priorities seems more limited than ever, as policies are embedded in the very logic of new institutions and as quasi-constitutional commitments. Changing course is made harder by the notion that renegotiating these policies would re-open the conditions of crisis they were meant to contain. The institutional landscape has been transformed, but broadly in the service of the existing socio-economic order. Extraordinary action has been for the sake of preservation, restoration, extension.

To grasp the workings of the EU in recent years, one must examine the ascendancy of a governing mode centred on the logic of emergency. This is the book's organizing thesis. The idea of extreme circumstances that need to be overcome, and that give licence for unconventional measures of last resort, has become central to how decisions are made. It has shaped the practice and legitimization of governing, the means by which authorities coordinate their actions and seek compliance. The book's aim is to unfold this logic, examining how its practices and ideas hold together. As I aim to show, with the concept of emergency rule one sees not just how a certain socio-economic order comes to be entrenched, but how the particular way this is done disables the democratic forces that might resist it.

Emergency rule is an old idea. In one form or another, it has been enacted and theorized from the ancient Roman Republic to the modern state. But it has now become fully cross-border in range, in the process acquiring a new set of dynamics. It is a distinctively *transnational* politics of emergency one sees in Europe more recently. This is true of the threats it is framed as a response to. Quasi-natural forces of the socio-economic sphere are cast in transnational terms, just like the things they are said to endanger. Indeed, officials highlight their cross-border character to underline the necessity of coming to terms with them. Unconfined in space, they threaten to be unconfined in their consequences. Their wide geographical extension becomes part of the rationale for governing from afar. The securitization of socio-economic affairs goes hand in hand with their scaling up to the transnational level.

Equally distinctive is the mode of governing itself. Whereas emergency rule has tended to be conceived as concentrating power in the hands of a unitary sovereign, here one sees the interplay of national governments with supranational agents of various kinds, working through the EU and around it. Decisions involve executive agents spread across multiple institutions and territories, without a clear hierarchical relation. Schmittian themes of sovereign exceptionalism have had a recent renaissance, but the target of our enquiry involves a significant reworking. This is an emergency politics informally co-produced by the many.

When scholars reach for the analytical frame of emergency and exception, some characteristic motivations tend to be in play. Three typical concerns come to mind, not all shared by this book. The first is a preoccupation with the rule of law, particularly how it can be maintained in extreme situations, and what is implied for the concept of the rule of law itself. Extraordinary moments become the test case for the limits of law. This book is not part of that genre, being political rather than legal in focus. Nor does it belong to a second body of work that sets its task as to define when emergency rule is legitimate. In the post-9/11 world, much scholarship has engaged in a normative project of defending the recourse to exceptionalism, from extraordinary rendition to torture. As will become clear, I have little such desire to rationalize the practice. Closer to the book's inspiration is a third tradition, in which the politics of emergency is approached as a practice of authoritarian rule, a window onto the ways in which democratic control is subverted or forestalled. Concerns of power and democracy are central to this book, with a focus on the challenges of executive usurpation. But I am interested also in what emergency rule can tell us about the EU's workings, and the light it can shed on European politics more widely. I approach the politics of emergency as problematic, in other words, but also as potentially instructive.

In the study of EU politics, the notion that challenging times can be the motor of change is commonplace. Indeed, that the EU has been built and strengthened through a series of crises is something of a disciplinary cliché. There is no need for a new book restating and glossing Jean Monnet's dictum 'that Europe would be built through crises, and that it would be the sum of their solutions'.[1] Such ideas have been thoroughly absorbed. But while they highlight the centrality of crisis politics in the international setting, they are limited in important ways. Perhaps the least of these is a tendency in EU scholarship to pose the matter in mechanistic terms, as a question of how breakdowns in integration are overcome. Crisis is approached as a moment in which the limits of a system are exposed, generating functional pressures for adaptation and innovation. Institutional change tends then to be read as a process of coming to terms with necessity, whether rooted in evolving socio-economic circumstances or the inadequacies of institutional design. Such accounts resemble those of public authorities themselves, who use them to rationalize their actions.

What gets marginalized are the political implications of what is done in the name of crisis response, as well as the very significance of measures being rationalized as such. A crisis suggests the breakdown of a whole, whereas typically

[1] Jean Monnet, *Memoirs*, trans. Richard Mayne (London: Collins, 1978), p. 417.

one is dealing with developments uneven in their effects, more critical from some material and ideological perspectives than others, and a true threat mainly for those deeply invested in the status quo. Actions always exceed mere functional adaptation, and that they are rationalized as crisis measures entails generalizing the perspective of some to be the perspective of all. One needs to recognize not just the intuitive point that crises can be instrumentalized to serve political ends, but that the very handling of situations as exceptional ones demanding actions of last resort is consequential in itself. By speaking of emergency politics, one brings these governing rationalities to the fore, and treats them as a self-standing phenomenon of investigation rather than the dependent effect of functional demands.

There is a second move to be made, important again for EU study. Amongst those avoiding a purely functional approach, there remains a tendency to see crisis politics as bound up in the specifics of particular policies and the shocks they are exposed to. The emphasis falls on what is particular to a given domain in a certain period. Interpretations of EU crisis politics that focus on the specifics of eurozone design are an example. Even as this can generate insightful analysis, it casts crisis politics as the symptom of a certain situation.

What is then overlooked is how the politics of emergency, as a wealth of wider scholarship confirms, constitutes a more general governing mode, independent of the specifics of a particular context. It is typically hard to localize in time: rather than the method of an isolated episode, it can be invoked recurrently, even indefinitely. Although the discourse may be about exceptional circumstances, temporary arrangements and extraordinary measures, one should avoid reproducing such assumptions in analysis: it is exactly ideas of the self-contained episode that may need to be questioned. Likewise, one should be attentive to the continuities across domains, as common factors encourage recourse to similar methods. As this book aims to show, in the practices of emergency rule one sees the crystallization of long-term tendencies irreducible to any one policy field. One sees part of a broader shift towards more coercive forms of rule under conditions of weakening democratic authority.

More particularly, the transnational politics of emergency marks, I suggest, the extent to which executive power has allowed itself to be reabsorbed into the socio-economic sphere, reversing the modern-democratic project of constructing political authority as an autonomous order. It is the culmination of a process of permeation and subordination, one that finds its highest form in the institutional arrangements of the transnational sphere—in the network of national, intergovernmental, supranational, and cross-cutting decision-making structures. Although European integration has often been advocated as a process of taming socio-economic forces, it has produced new sites of exposure and adaptation. In place of the aspiration to set priorities and weigh normative claims, executive power has come to embrace a reactive approach centred on adjusting to the force of presumed necessity. The authority of political and legal norms becomes conditional on their compatibility with socio-economic demands.

Whereas the emergence of the modern democratic state was widely conceived and rationalized as the de-personalization of power and its regularization, decision-making today is increasingly relocated to ad hoc settings that escape codification and depend on individual discretion. The informalization of power is widespread. The transnational politics of emergency is the context in which these tendencies find their most concentrated expression, as well as a key context for efforts to rationalize and legitimize them.

The major theories of EU politics, still variations on intergovernmentalism and functionalism, are generally institutionalist accounts. In one way or another, they focus on the roles, rules and interests of institutional agents, taken as differentiated and relatively enduring. What they underplay are the informal, cross-institutional ties that emergency rule brings forward. Most at ease when examining the agendas pursued by states or supranational institutions, they are less equipped to discuss what cuts across these formations—the ideological and socio-economic influences common to a wide array of executive agents, or the shared claims to authority they make. Emergency rule sees institutional structures bypassed or reshaped in the image of these networks and demands. Beyond mere redistributions of power across distinct and stable entities—the Commission, Council, states, and so on—one sees shape-shifting structures and makeshift arrangements like the Troika that are ambiguous in status and uncertain in duration. Power is personalized, and detached from the bureaucratic routines that support talk of institutional agendas and interests.

Disputes over sovereignty are one of the leitmotifs of emergency politics. When existing norms are broken, the inevitable question will be whether those doing so were entitled to. Did they have the authority to identify the circumstances as exceptional and to prescribe an extraordinary response? In recent European politics one sees renewed efforts to answer with certainty the sovereignty question that has tended to be avoided in the EU. Emergency politics returns it to the table. But while the question may be posed with renewed force, it is a misleading one in current conditions if one seeks answers in the form of a pre-eminent institution. Executive power has become blurred to the extent that political exceptionalism is a joint endeavour.

When first writing on this topic in the early 2010s,[2] I focused on measures taken in an especially intense phase of eurozone crisis politics—between the first Greek bailout and the making of the Fiscal Compact. While this remains an exemplary case of emergency politics in the EU, one to which this book continues to refer, the passing of time makes a broader view possible. The intention is to step back to

[2] Jonathan White, 'Emergency Europe', *Political Studies* 63 (2) (2015), pp. 300–18.

consider the longer-term factors shaping these practices, as well as some of the legacies arising. Emergency rule marks the radicalization of existing tendencies, but it also leaves its own distinct traces. It points not just to a set of institutional transformations but to a potentially enduring recalibration of authority. Despite the emphasis on exceptionality, emergency rule sets precedents for what follows. Notable is not just how extreme conditions come to be invoked at a particular moment but how they become a reference-point for future actions. When the Italian President made his provocative intervention in May 2018 in the ministerial appointment process for the newly formed Lega/Movimento 5 Stelle (M5S) government, it was with explicit reference to the threat of renewed uncertainty on the financial markets: the language of 'the spread' that had dominated Italian public discourse in the autumn of 2011 was pointedly emphasized once more.[3] A period of emergency, however purportedly unique and self-contained, generates echoes that are heard down the years. The danger of a relapse into yesterday's crisis, and the need to ward off tomorrow's, become persistent themes in the discourse of power. The effect is that ideals of authority grounded in democracy, technocracy, and proceduralism are challenged not just within the bounds of a delimited episode but in more open-ended and lasting ways.

There is another way that emergency politics leaves its mark: in shaping the political response it meets. Part of the commonsense of recent political commentary has been that a rebellion is unfolding against the policies and authorities of the liberal order, typically framed with the category of *populism*. As this book argues, such diagnoses typically miss something crucial, overlooking how a central strand in these counter-movements is exactly the repudiation of emergency rule, i.e. of extraordinary decisions rationalized by appeal to necessity. As we shall examine, what goes by the name of populism tends to be not just the critique of a failure of policy, nor the condemnation of detached and corrupt elites, but of the genuflection of authorities before ramped-up socio-economic demands. It is a critique of political acquiescence.

While emergency rule entails frenetic decision-making, its decisions are rationalized as unchosen and unavoidable both in substance and timing. To anticipate a distinction we shall come to, it is characterized by heightened executive activity—or '*doing*', as I term it—coupled with heightened disavowals of *agency*, i.e. the capacity to choose freely between options. Emergency rule responds to forces it casts as all-powerful. What tends to be called populism is in significant part a repudiation of the discourse and performance of necessity that emergency politics puts centre-stage. Here one sees how the concept of emergency is valuable not just for understanding executive power in the EU, but also for grasping its critique.

[3] For Mattarella's own account of his actions: http://www.quirinale.it/elementi/1417; on the new concept of 'lo spread' in Italian public discourse: https://www.ilpost.it/2015/03/07/storia-spread/; also Chapter 3.

Although in many ways a very British affair, in the currents of Brexit such themes converge. With its denunciations of 'Project Fear', the 2016 Leave campaign was, amongst other things, a rejection of the politics of necessity and a call for renewed agency. It defined itself by its hostility to the ideas of economic emergency put forward by forecasters in the Treasury, IMF, EU institutions, and the markets. It was, by its own account, an effort to reassert the primacy of politics over economics, identifying this as a project for the sovereign nation-state acting in contradistinction to the dynamics of the transnational sphere.

Yet at various points over the following years, it seemed the best hope of delivering Brexit in any form would involve relying on the same themes of compulsion and urgency the Leave campaign had condemned. From the winter of 2018/19, the prospect of market turmoil should the Withdrawal Agreement be voted down was raised by central bankers and political strategists to focus minds in Parliament.[4] The EU's many veto players were long highlighted by the UK government to emphasize the difficulties of renegotiation, and reiterated by EU representatives equally keen to see a deal concluded. Postponing votes and debates in Parliament, the Prime Minister's circle repeatedly sought more time to highlight the stakes of 'no deal'. MPs were cast as obstacles from all sides, and the likelihood of chaos and emergency rule was openly invoked by representatives of the UK and the EU alike.[5]

In the manner of a self-cancelling prophecy, the prospect of disorder was raised as a way to pursue order, to encourage all but the most reckless to fall in line. In the manner of a self-*fulfilling* prophecy, the same strategy provoked further anxiety within markets, business groups, trade unions, and EU authorities, giving new fodder for the politics of emergency.[6] For all its specificities and unintended outcomes in this context, the pattern of genuine threats amplified by the tactics of a government was a local reprise of manoeuvres seen repeatedly in Europe's crisis politics of the 2010s. Leading figures in the executive sought to sustain and even bolster themselves in adversity by drawing on the logic of exception.

A larger theme we shall explore is how rejoinders to emergency politics may reproduce some of its defining characteristics. Efforts to challenge its political legacy may include elements of exceptionalism themselves—both when agents seek to exit the EU and when seeking to contest it from within. One sees a

[4] https://www.ft.com/content/702e5dbc-ed8d-11e8-89c8-d36339d835c0

[5] See e.g. Caroline Wheeler, 'UK ready to declare martial law to avert no-deal chaos', *Sunday Times*, 27 January 2019: https://www.thetimes.co.uk/edition/news/uk-ready-to-declare-martial-law-to-avert-no-deal-brexit-chaos-bfqgzzlrw; also PM May's notorious speech of 20 March 2019 (https://www.theguardian.com/politics/2019/mar/20/acting-like-trump-theresa-may-sparks-mps-brexit-fury) and the wilder acrobatics of the Johnson government. On the EU side, see the Commission's proposals of 19 December 2018: https://ec.europa.eu/info/sites/info/files/com-2018-890-final.pdf

[6] See e.g. the joint letter of the UK Confederation of British Industry (CBI) and Trades Union Congress (TUC) to PM May of 21 March 2019, warning of a 'national emergency' should the Government keep to its course: https: //www.theguardian.com/politics/2019/mar/21/heads-of-tuc-and-cbi-write-to-may-pleading-for-a-brexit-plan-b

counter-politics of emergency, emerging 'from below' in parties and movements, replete with its own set of normative challenges. This is where executive discretion meets the practices of disobedience. How a defensible rejoinder to the transnational politics of emergency might look is one of the guiding questions for the book's later chapters.

Much of the recent scholarship on emergency rule displays a liberal concern for its effects on individual rights, notably those of assembly and arrest. Such focus is in keeping with a concern for the rule of law. Yet the implications of emergency rule for democracy are no less pronounced, especially when the phenomenon is transnational in scope and the range of decision-makers extends well beyond those elected by the population affected. Emergency rule relies on and entrenches asymmetries between executives on the one hand, be they in governments, supranational institutions or functional agencies, and those who would scrutinize them on the other. It is the political implications of emergency rule that are my focus.

There are those for whom executive pre-eminence and decision are part of a positive story to be told of crisis management. The rise of the 'executive unbound' is admired by some as the rise of an agent that can get things done, more capably than legislatures and lower-tier bureaucracies.[7] The argument has been powerfully advanced in the US, and shows signs of being adopted in Europe too. Taking stock of the EU's recent transformations, and the role of the European Council especially, Luuk van Middelaar endorses the 'emancipation of the executive' he sees in progress, as a way to build strength and global standing.[8] Such transformations mark the polity's coming of age, its increasing self-awareness and governing capacity. Political authority consolidates in the encounter with pressing 'events', albeit some eggs may be broken along the way. As will become clear, I am sceptical. Even when elected officials are to the fore, the democratic costs tend to be high, and not just in the short and medium term. More than just improvisations, the EU has been witness to suspensions and evasions, and these leave a lasting legacy. Moreover, the executive forms emerging remain closely tied to defined policy goals, however emancipated from law and procedure they may be. They are committed to the reinforcement of existing socio-economic structures, and thus in an important sense mark continuity rather than new beginnings.

This is a concept-driven book. Its main contribution is intended to lie in the concepts its advances and the connections suggested. Rather than to present an historical account of the recent past, its aim is to explore the logics in play and the terms best suited to grasping them. *Emergency* stands as the master concept, but

[7] Eric A. Posner and Adrian Vermeule, *The Executive Unbound: After the Madisonian Republic* (Oxford: OUP, 2011).

[8] Luuk van Middelaar, *Alarums and Excursions: Improvising Politics on the European Stage* (Newcastle: Agenda, 2019), pp. 175–6.

the following chapters put forward a range of others, analytical and normative, intended to clarify contemporary practice and its organizing ideas. Amongst those we shall become familiar with are *de-institutionalization*, *crisis expertise*, and *constitutional reassurance*. A number of heuristic dichotomies are proposed, including those of *know-that* and *know-how*, of *agency* and *doing*. In addition, some of the concepts of contemporary discourse that are most familiar—democracy, technocracy, sovereignty, authority, and populism—will be placed in a distinctive light.

Conceptual work of this kind relies on moving back and forth between theoretical reasoning and empirical evidence, using the latter to refine the former. If the outcome is better insight into the context at hand then the exercise fulfils its primary goal. But the hope is that the categories of analysis developed are transferable to other contexts and can be used to draw connections across them. Even as the EU's emergency politics of the 2010s recedes from the immediate present, some of the insights it can yield should retain their broader significance. The EU in this period stands as an exemplary case of a kind of emergency rule too rarely examined: socio-economic rather than geo-military in focus, transnational and dispersed rather than statist in range, and the means by which not just individual rights but norms of collective self-determination are cancelled or forestalled.

In the book's first chapter, I outline some key features of the politics of emergency, its distinguishing elements in the transnational context, and the value of this analytical frame. The account is built on an analysis of the EU's handling of economic malaise in the 2010s. Central to the political imaginary in these years was the figure of the Troika. Though on one level just a term of public condemnation, on another it encapsulated some deeper truths about the politics of the time. Specifically, the Troika concept expressed a politics based on ostensibly impermanent, unconventional arrangements for defined purposes, governed by an agenda of speed and urgency, and propelled by trans-institutional executive power. Taken together, these features amount to a governing mode that theories of emergency rule can illuminate. They help us to understand how such features interrelate, and draw attention to aspects under-emphasized in current scholarship, including the shape-shifting of institutions and the networks that cut across them. Equally, classical takes on the sovereign state of exception can be a foil by which to identify the distinctive aspects of EU exceptionalism, including its informal character and its co-production by many actors. The chapter acts as a point of departure for thinking about the specificities of transnational emergency rule.

As Chapter 2 argues, certain structural features of the EU create an in-built vulnerability to discretion under the sign of emergency. Some of the basic drivers are long-standing, even if their full implications become visible only in the context of more recent changes. An important feature is how the EU order has been constructed to serve particular purposes. Since its origins in the 1950s, European integration has entailed the creation of institutions whose rationale is to advance and maintain certain policy ends—notably those underpinning the 'four freedoms' of the common market, understood as ways to advance broader goals of peace and prosperity. As the chapter argues, the effect is that policy commitments have been privileged over procedural arrangements. Rather than self-standing entities that can be put to different ends, broadly on the model of the modern state, one sees institutions evolving with the policies, and liable to be side-stepped should they fail to serve them. There is an instrumentalist orientation to the transnational polity that makes its arrangements precarious when the going gets tough. Moreover, a non-hierarchical constitutional structure does little to inhibit these restructurings, indeed arguably gives further encouragement. A long-standing idea in constitutional thought is that the diffusion of power across many institutions can create political paralysis, especially when the division of roles is weakly underpinned constitutionally. Arriving at common policies, enforcing and revising them is the work of many hands, always potentially convoluted. The ideas and practices of emergency are a way to galvanize action, coordination, and innovation across a diverse and potentially recalcitrant institutional field.

Developments in recent decades have pushed the EU from a structural vulnerability to emergency rule to an increasing reliance on it. Susceptibilities once latent have become increasingly manifest. As Chapter 3 shows, this happens in the context of a series of challenges to political authority. By the turn of the millennium, leaders found themselves surrounded by powerful non-state actors claiming the authority to interpret socio-economic conditions, to make sense of moments of uncertainty, and to specify the responses they demand and when. Market analysts and ratings agencies are emblematic examples of these emergency interpreters of the private sphere. Structurally attentive to conditions of volatility and stress, such agents foster an atmosphere conducive to the politics of emergency, and—as those often invested in maintaining and reinforcing aspects of the status quo—concertedly influence the considerations elevated to the status of a threat. Moreover, due to changes within the field of executive power itself, the voice of these agents carries ever further into decision-making circles. The rise of *governance* marks the rise of an executive ethos characterized on the one hand by the decisive prioritizing of problem-solving capacity over politico-legal norms, and on the other by the search for public consent in mechanisms other than responsiveness to public will. The first reinforces a willingness to escape procedural constraints, while the second encourages the serial invocation of exceptional circumstances as a way to give citizens reason to accept unpalatable decisions they were left out of. The politics of emergency is a way of coping with weakening public authority in the age of governance.

The events of the 'Euro crisis', from the handling of market pressures to the ECB's reincarnation as lender of last resort, are just one manifestation of this governing pattern. The chapter goes on to examine how these dynamics, rather than exclusive to one scenario or policy area, play out across a wider field of transnational decision-making. The handling of economic malaise is paired with the political response to migration in the years after 2015, with the suggestion that both are the consequence of executive deference to non-state power and a governance approach that seeks validation in fast action.

Emergency politics has become recurrent in the EU. Interestingly, though, it tends to be coupled with the suggestion that it will soon be consigned to the past. Notions of a 'normality-to-come' are part of its legitimizing apparatus, and a plausible aim if the rhetoric of emergency is not to be devalued by overuse. Yet such notions of imminent normality can be little more than an ideological claim unless the emergency regime can be credibly terminated. Once embarked on emergency rule, executive agents face questions of whether, how and when they will end it. Can they draw a line such that their unconventional measures are seen to be contained: can they provide *constitutional reassurance*?

Chapter 4 analyses the preconditions for an act of separation between exceptionalism and its aftermath. Given the informal character of EU emergency rule, this is a political rather than legal question, centred on credible claims to the effect that crisis has been superseded, or that executive power is newly constrained, or that it is now peopled by those less inclined to exceptionalism. The chapter articulates and explores these possibilities. It questions whether the resources for demarcating emergency rule from its aftermath are available. Partly a matter of the amorphousness and uncertainty of the extreme circumstances cited, it is also a function of the EU's structure. Features of emergency rule will tend to endure, with important implications for the kind of authority that can be accommodated. Procedural authority is hard to renew, even for those who seek it, and its likely weakness one of the legacies of emergency rule. Through the prism of the EU's constitutional particularities, the chapter addresses an under-examined aspect of exceptionalism more generally—the constitutional problems arising from the difficulty of localizing it in time.

When procedural authority is persistently in doubt, one can expect the invocation of other forms of authority. Given its long-standing association with EU politics, especially relevant is that of *technocracy*. On one view, emergency rule spells problems for expert-led rule, challenging existing knowledge and encouraging the intrusion of non-scientific criteria on decision-making. Especially when crisis management forces collaboration between multiple institutions, the technocratic credentials of each come under strain. Yet as Chapter 5 tries to show, emergency and technocracy are also complementary, and some may embrace exceptionalism precisely to promote technocratic authority. Knowing how to act in atypical situations is in some ways the very measure of expertise. Only in irregular times can the claim to expert status be renewed. Examining recent EU experience, the chapter shows how claims to expertise can span theoretical and practical knowledge. Emergency rule disrupts claims to scientific 'know-that', and encourages would-be technocrats to rely on a form of *crisis expertise* centred on practical know-how. Ideals of *techne* give way to *phronesis*. Rather than the termination of technocratic logic, emergency rule signifies its redefinition—and in politically problematic ways. This recalibration facilitates the diffusion of 'technocrats' across different institutions, including those meant to be democratic, and leads to a precarious form of authority resting on weakly underpinned trust. The analysis contributes both to our understanding of the EU and how to think about technocracy more generally.

Questions of authority raise questions of public consent. Here the book moves to examine rejoinders to emergency rule: the critical responses it generates and, more normatively, those one might wish to see. I proceed from the observation that the politics of emergency, with its suggestion of decisions born of necessity rather than choice, consolidates an image of the EU as the negation of political agency. Chapter 6 examines how such actions invite their opposite: a politics defined by its rejection of necessity and the celebration of political *volition*. Here one sees another of the longer-term legacies of emergency rule, and in particular how it brings sovereignty themes to the fore (think Brexit). The mobilizations across Europe commonly studied under the heading of 'populism' gather, I suggest, much of their wider appeal from the promise to restore political agency—a promise independent of the other features typically ascribed to them, and which the term populism can serve to obscure. With the distinction between the politics of necessity and volition, one can better understand the motivations in play, and avoid casually attributing the success of such movements to their least desirable features (nativism, anti-politics, and the like). Further, one gains insight into the reasons why established executive powers may be moved to respond harshly—so as to protect the securitization moves on which emergency rule depends. The chapter's final section examines how the spectre of such counter-mobilizations may be invoked by established powers as reason for further constraining measures. This raises the unwelcome prospect of escalation—but also a persistent demand for agency that could be harnessed for democratic ends.

Efforts to stabilize the political scene and to ward off the next crisis typically involve moves to restrict the discretion of all but the most powerful. Attempts to lock in the policy regimes of the new order are a characteristic feature of emergency rule. Besides reinforcing the appeal of a voluntarist stance, one implication is that those wishing to reshape that order may be compelled to do so by rule-breaking means. But if one rightly has concerns about the unbound character of emergency rule, what is one to make of the kinds of transgression that may be required to overturn its legacy? Chapter 7 examines these practices of resistance using the concept of *disobedience*. From the Syriza-led government in Greece to the Movimento 5 Stelle/Lega government in Italy, one has seen the appearance of parties and movements stating a willingness to challenge the rules of the post-emergency EU. The contours of an extraordinary politics 'from below' start to emerge, as dissenters seek to reform or unmake the status quo.

The concept of disobedience offers both an interpretative frame for these mobilizations and the basis for their evaluation. One of the chapter's purposes is to develop criteria by which to distinguish principled from illegitimate forms of extra-legality, and to connect them to developments on the ground. As a contribution to the political theory of civil disobedience, it reflects on the kinds of agent that can lead it, showing how one of the distinctive features of the transnational realm is to enable a form of disobedience led by parties. In addition to the activities of national parties, some of the possibilities are foreshadowed in pan-European groups such as the democracy movement DiEM25. With the concept of principled disobedience, one has the basis on which to understand what motivates such responses to emergency rule, as well as to take critical stock. The chapter concludes with some reflections on the normativity of exceptionalism more generally.

Is there the prospect of an escape from the politics and counter-politics of emergency in today's Europe? Disobedience marks an incomplete break with the logic of exception, so how might a more thorough rejoinder look? This question is taken up in the penultimate chapter. In the nation-state setting, answers are typically sought in procedural checks on executive discretion, whether in the form of legal constraints upheld by courts, or strategies of institutional design that make emergency rule less appealing. There is much to be said for such moves in the EU, and proposals exist of how the EU might benefit from a constitutional reorganization of this kind. But as the chapter argues, what tends to be missing is an account of the political will that could support such initiatives. This matters because procedural remedies are likely to depend on the approval of those most likely to block them—the established agents of executive power. Moreover, even once instituted, such changes need to be maintained, given the range of socio-economic and political forces ranged against them.

The chapter develops an account of the political restraint of emergency rule based on mobilization within broadly existing structures. Only, I suggest, the re-embedding of decision-making power in a world of ideas-based collectives can realistically temper the tendencies at hand. With emergency politics symptomatic of the hollowing of parties, challenging it entails strengthening the ties of partisanship, both as these may constrain directly the discretion of representatives holding authority, and as they may influence it from a position of opposition. The chapter draws and develops on debates about transnational partisanship, connecting them to a discussion of the structural preconditions of opposition. Emphasized is the significance of temporal features. The experience of modern democracy suggests that the periodicity of institutions has been an important underpinning for critical opinion. Associations come alive by virtue of their rhythms, and ideologically strong and enduring modes of opposition depend on recurring moments of visibility and conflict. By contesting elections at all political levels, but most especially by contesting *national* elections as *transnational* partisans, those critical of executive discretion under the sign of emergency may take steps towards its political constraint.

But there is a deeper significance too to the rhythms of partisanship, to do with the demarcation of political from socio-economic power. With its urgency of word and deed, emergency rule signifies the subordination of political decision-making to the temporality of the socio-economic sphere. One of the contributions of institutions of periodicity in the state context has been to mark out the autonomy of political time. By establishing time-structures independent of the passage of events, such institutions offer a framework for the coordination of opinion at the level of general principle, abstracted from the specificities of a given context. If the EU is to be more than the sum of ad hoc responses to socio-economic forces, this kind of demarcation seems crucial. Here lie some of the resources by which to rebuild the autonomy of executive power and make it susceptible to democratic control.

The concluding chapter offers a restatement of why the politics of emergency is something one should wish to see contained. It highlights a feature that reinforces one of the problems of the EU's political culture: the tendency to frame events as a series of one-off, sui generis episodes, each detached from a larger historical and normative frame. With their emphasis on exceptional measures for exceptional times, the protagonists of emergency rule assert the extraordinary nature of the situation they face. Emphasizing that their actions are of last resort, they claim to be bound by no precedent and to set no precedent. Theirs is a politics of discretion grounded in the supposed particularity of the moment: a politics of the *singular*. As the chapter argues, the relation between democracy and the unique is double-edged. On the one hand, these are the conditions of innovation, experiment, and creativity in politics. Discretion in unfamiliar circumstances is what enables a polity to evolve, a thought expressed in the narratives of integration-through-crisis that have tracked the EU's history. But there is a darker side to the politics of singularities. Without the recognition of continuities and the insistence on standards and principles that bridge events in their specificity, governing agents are cut too much slack. The critical scrutiny of power and the identification with its demands depend on the recognition of what is *not* particular to the moment—on moving beyond the politics of singularities.

As the book aims to show, the transnational politics of emergency powerfully shapes how institutions and policies have come to be made, inside the EU and alongside it. A world of institutional discontinuity and executive discretion co-exists with, and helps support, a socio-economic order relatively enduring until now in its essentials. Although the logic of emergency evokes exceptional departures from a stable framework, it should be studied not just as the governing rationality of particular episodes but as something that selectively locks in the past and shapes anticipation of the future. Although it has become customary to treat the EU as afflicted by a succession of unique challenges, ranging across money, migration, geopolitics, and more, one should look for the continuities that span the domains.

Perspectives on the EU are often binary in form—for or against, leave or remain—and one effect of the Brexit process has been to accentuate this. In Britain and beyond, critical analysis of the EU has become associated with calls for exit, dissolution, or predictions of demise. At the same time, one of the more unfortunate legacies of the Brexit story may turn out to be strengthening the opposing sentiment: an undiscriminating attachment to the European project across sizeable sections of the EU public. Sympathetic critique is side-lined in both cases. This book avoids dismissing the EU, either as destined to fail or as something whose problems can be left to others. But it also aims to highlight some deep-seated flaws, to explore how EU structures can magnify certain problems and entwine them with dynamics that are distinctive to the transnational sphere.

While, for many, European integration has been a process of constraining sovereignty and the wayward exercise of executive power, the effect it seems has been to create the conditions for a new kind of political exceptionalism, different in structure but no less problematic. It is being contested in today's Europe in ways that may limit it, but the prospect of its resurgence will not go away. Deeper than any one set of institutional forms, there is little reason to suppose even the disintegration of the EU would bring the politics of emergency to an end. Indeed, were the EU to dissolve or splinter, the arising conditions would surely favour further appeals to executive discretion under the sign of emergency. Emergency Europe would persist. This is a mode of rule difficult to consign to the past, even as determined agents seek ways to contain it. It is worth getting familiar with its forms and rationalities.

1

Beyond a Sovereign State of Exception

In the political imagery of Europe in the 2010s, a central figure was the *Troika*. This was the name given to a group of institutions charged with restoring order to a turbulent economy. Composed of representatives of the European Commission, ECB, and IMF, its role was to build stability by restructuring the policies of states that had taken on loans. 'The Troika is not a formal organisation', its functionaries observed. 'It is the result of the close cooperation between three different institutions [...whose] respective focus and expertise complement each other.'[1] Its name would become both a term of self-description and of popular condemnation.

New political forms are often captured by metaphors, and the figure of the Troika was rich in association. As commentators would soon observe, it described in its original meaning something far from the world of political institutions, though not without features of resonance. A troika was a chariot drawn by three horses, assembled for the purpose of speed. Appearing in early-modern Russia, it was a vehicle out of the ordinary, for pressing tasks such as the relay of information. It had symbolic qualities too as a model of cooperation. The troika form expressed the coordination of entities distinct yet in a complementary relation. It would become a staple theme of Russian painting, as an image of order and power, sometimes of ceremony.[2] The horses of the troika would draw the eye, but the scene extended to the chariot behind, where a human figure sought to steer the arrangement and maintain its forward motion. Completing the scene was a landscape, largely inanimate, waiting to bear the vehicle's imprint.

The figure of the troika would reappear in later Russian history in a darker guise. In the Soviet Union it described an instrument of repression. The troika was an extra-judicial three-man committee, used in many regions and in more than one period, whose role was to issue punishments for dissidents.[3] Operating outside the usual legal framework, a troika used extreme methods to reach its

[1] 'Responses to Written Questions on the Economic and Social Crisis in Greece by Several Political Groups within the European Parliament', by Olli Rehn (Commission Vice President) and Jörg Asmussen (Member of the ECB Executive Board), p. 2.

[2] Nicola Kozicharow, 'Filipp Maliavin's Troika Dream of 1933', paper presented at Cambridge University, 15 February 2011.

[3] Michael Ellman, 'Regional Influences on the Formulation and Implementation of NKVD Order 00447', *Europe-Asia Studies,* 62 (6) (2010), p. 917. Troikas were composed of a regional NKVD chief, a party leader and prosecutor; they were used in the liquidation of kulaks in early 1930s and again in the Great Terror of 1937–8.

Politics of Last Resort: Governing by Emergency in the European Union. Jonathan White, Oxford University Press (2020). © Jonathan White.
DOI: 10.1093/oso/9780198791720.001.0001

verdicts, generally in secret and without possibility of appeal. Its purpose was to accelerate the sentencing process.[4] For these groups of local officials, the task was to execute Moscow's wishes on the ground, quickly, efficiently, and as best they saw fit.

As a guide to the European political scene in the 2010s, in many ways the imagery of the troika is misleading. Amongst the more dubious implications is that of a clear definition of roles. Between the representatives of the European Troika there were uncertain and ill-defined ties, and occasionally some important disagreements. The contours of the formation were not always clear. Opaque too were the connections to those steering the Troika, generally assumed to be the representatives of states sitting together as the European Council, the more powerful amongst them in control.[5] Relations with the Council could be fractious, and hardly based on a strict separation of kinds. Politics and technocracy, agent and principal, were not so easily distinguished. Nor was the cast of actors settled. There was a shape-shifting quality to the arrangements of the period, not well captured by the Troika name.[6] And while the actions of the ECB, Commission, and IMF were influenced by the directions of the Council, they also acquired a dynamic of their own. Sometimes the chariot was not in command.

But even if the idea of the Troika is not wholly reliable, and in any case but one part of the story, it captures some important truths of the period. It is a revealing stock of metaphors for the times. It expresses, I suggest, three points in particular, to be developed in the following sections. First, it conveys the idea of an unconventional arrangement, set up for specific ends. The Troika was a formation convened for a purpose. Second, it conveys the idea of an arrangement governed by an agenda of speed, responsive to notions of urgency and the unusual. It was for occasions out of the ordinary. And third, it speaks to the centrality of executive power, yet a power divided across multiple institutions, whose representatives were faced with the ongoing challenge of maintaining their precarious coordination.

These elements add up to a distinctive format, one I shall argue can be grasped as a form of emergency rule. Such a format cuts across the conventional picture of the EU, in which EU politics is shaped primarily by the interplay of distinct institutional interests and agendas. In a context of rapid institutional change, we need to recalibrate how we look at the EU. Rather than focusing on the usual institutions and their changing fortunes—the relative gains of the Commission and Council—one should examine the emergency logic by which new formations

[4] Timothy Snyder, *Bloodlands: Europe Between Hitler and Stalin* (New York: Basic Books, 2010), pp. 26, 82.

[5] On Troika powers: C. Randall Henning, *Tangled Governance: International Regime Complexity, the Troika, and the Euro Crisis* (Oxford: Oxford University Press, 2017).

[6] The Troika would be expanded in 2015 with the addition of the ESM; the IMF's withdrawal was meanwhile an ongoing possibility.

emerge and old ones are straddled and reconfigured. What needs study is a particular mode of rule and its guiding rationality, for which temporary formations such as the Troika may be a better guide than the long-familiar structures. Ultimately short-lived as these creations may be, they embody some lasting truths, and produce effects that are lasting too. Though talk of the Troika would later recede, its basic features had an enduring significance.

In reconstructing EU politics in these terms, I intend also to show how theories of emergency government may need to be adapted in today's world. Formations such as the European Troika, as a compound of agents external to the state yet rather less than a supranational executive, suggest a model of extraordinary rule quite different from a sovereign state of exception. Emergency rule should be decoupled from its associations with the sovereign figure, reversing the Schmittian move. Equally, its logic of security needs to be recognized in fields far removed from the context of open war, in a world of technocracy, economics, and private authority. The Troika can be viewed as an expression in miniature of some more general patterns in the transnational politics of emergency.

Escaping Norms

A defining feature of the Troika arrangement, and analogous creations of the period, was their uneasy relation to the pre-existing EU institutional field. Neither quite of the EU nor self-standing arrangements, they fitted awkwardly with the procedures of the day. They were novel and impromptu entities. Though not necessarily at odds with EU or national law, often they were anticipated in neither. The Troika had no basis in the EU treaties, and little in the way of an agreed structure. It brought together institutions whose established roles bore little resemblance to the duties of monitoring and enforcement expected of them in this context.[7] Here was a form of politics that escaped the hold of established procedures in national and EU politics, and of the wider norms typically associated with them.

The Troika was emblematic but not unique. It was flanked by a range of 'special purpose vehicles' created from scratch to achieve defined ends, emergency lending facilities prominent amongst them. Reliance on the unscripted was visible in one of the main decision-making forums of the period, the Eurogroup, where anything like a defined operating code was deliberately eschewed. As the Greek Finance Minister was reportedly told by the Eurogroup's secretariat at a critical meeting in summer 2015: 'the Eurogroup does not exist in law, as it is not part of any of the EU treaties. It is an informal group of the finance ministers of the Eurozone

[7] Henning (*Tangled Governance*), p. 8. Elements of the Troika format had already been trialled in Romania from 2009, involving the IMF, World Bank, EU, and EBRD.

member states. Thus there are no written rules about the way it conducts its business, and therefore its president is not legally bound.'[8]

Much has been written on how actions in the Euro crisis[9] challenged specific provisions in EU law.[10] There is no need to comb over the details once more: it is enough to highlight the pattern. One saw a range of measures departing from established standards, in the sense that these were overtly transgressed or that their application was suspended. It is important here to distinguish positive rules from basic norms, for not all transgressions may be equally significant. In any imperfect order there may be codified rules that cannot reasonably be upheld at all times, and some that may deserve to be abandoned. Amongst these might be rules of policy designed for a different age, legitimately set aside by decision-makers exercising discretion. But whereas improvisations of this kind might be applauded, there are deeper commitments less lightly waived—things that underpin politics rather than limit it. More significant in this period than the breaking of certain policy rules was the infringement of some of the fundamental norms of modern politics and of the institutional procedures that plausibly serve them.

I have in mind certain widely accepted standards of appropriate conduct as these apply to both EU and national politics. Norms link ideas of right action to procedural scripts held to express them.[11] They combine ideas of how things are with ideas of how they should be, and while they may be imperfectly rendered in existing practice, they have established associations with concrete forms of conduct such that they are not wholly abstract or ideal. While some may have a textual basis, being commonly referred to as laws, others may be mainly tacit, in the manner of Dicey's 'constitutional conventions'.[12] These standards, we may stipulate, are the subject of broad consensus, even if constantly evolving and persistently subject in some measure to disagreements of interpretation.

[8] Yanis Varoufakis, *Adults in the Room: My Battle with Europe's Deep Establishment* (London: Penguin, 2017), p. 447, reporting Jeroen Djisselbloem's words at the Eurogroup meeting of 27 June 2015.

[9] The term is of course contentious, denoting not only certain events but tying their explanation to Eurozone design. I use it as a placeholder, using more focused terms where needed.

[10] The consistency of the emergency credit facilities with the 'no bail-out clause' of Article 125 (1) of the Treaty on the Functioning of the European Union (TFEU), and the legality of the ECB's 'Outright Monetary Transactions' (OMT) policy in light of Article 123, have been the main focus: see e.g. Kaarlo Tuori and Klaus Tuori, *The Eurozone Crisis: A Constitutional Analysis* (Cambridge: Cambridge University Press, 2014), chap. 5; Claire Kilpatrick, 'On the Rule of Law and Economic Emergency: The Degradation of Basic Legal Values in Europe's Bailout', *Oxford Journal of Legal Studies* 35 (2015), pp. 325–53; Christian Joerges, 'Law and Politics in Europe's Crisis: On the History of the Impact of an Unfortunate Configuration', *Constellations* 21 (2014); Thomas Beukers, Bruno de Witte, and Claire Kilpatrick (eds), *Constitutional Change through Euro-Crisis Law* (Cambridge: Cambridge University Press, 2017), pp. 15ff.

[11] Cf. Geoffrey Brennan, Lina Eriksson, Robert Goodin, and Nicholas Southwood, *Explaining Norms* (Oxford: Oxford University Press, 2013), esp. chaps 1–2. NB norms are not to be equated with settled practice, since they can persist in the breach, and given not all behavioural regularities are norm-guided.

[12] For an overview: Geoffrey Marshall, *Constitutional Conventions: The Rules and Forms of Political Accountability:* (Oxford: Oxford University Press, 1987).

The *equality of states* has long been one such norm of the EU. Honouring it has meant governing interstate relations with prescriptions of general application, including those of EU law, rather than contracts and commands targeted at particular states. In the EU's crisis politics of the 2010s, this norm was repeatedly transgressed. The existence of the Troika, with its country-specific demands aimed at debtor states, was itself a significant departure. Its effects were reinforced by parallel moves departing from the principle of unanimity in treaty-making and amendment. One saw treaties targeting particular states and institutions made outside the EU framework to escape the constraints of unanimity.[13] While consistent perhaps with international law, even with EU law, such practices broke with the expectations and conventions of treaty-making in the EU, and suggested that what mattered was the consent of just some.[14]

State sovereignty on matters not conferred to supranational institutions was another such EU norm. Although European integration has always posed questions of sovereignty, compounded by wider global developments since the Cold War, a widely endorsed standard it remained. It found backing in treaty sources,[15] and was conventionally expressed in nationalized procedures of decision-making for taxation and welfare policy. The Troika's use of conditional loans to enforce cuts in the public expenditure of certain member-states was just one development at odds with it. The degree of pressure brought to bear on state governments would later be revealed in private letters sent by the ECB head detailing fiscal policies to be pursued in return for access to central-bank funds.[16] Such moves were matched by broader efforts to make national budgets open to supranational inspection and revision, again departing from an established understanding of the Union's division of powers.[17] External involvement in key national elections, and the associated appearance of technocratic governments in Italy and Greece in autumn 2011, were likewise challenges to state sovereignty.

[13] See e.g. the Fiscal Compact and EFSF: Brigid Laffan and Pierre Schlosser, 'Public Finances in Europe: Fortifying EU Economic Governance in the Shadow of the Crisis', *Journal of European Integration* 38 (3) (2016).

[14] For a useful overview of the measures adopted in different countries, see the reports available at eurocrisislaw.eui.eu. For a statement concerning the equality of states see Article 4(2) TEU.

[15] See Article 5(2) TEU: 'Under the principle of conferral, the Union shall act only within the limits of the competences conferred upon it by the Member States in the Treaties to attain the objectives set out therein. Competences not conferred upon the Union in the Treaties remain with the Member States.'

[16] See the letters sent by Trichet to Ireland (19 November 2010), Spain (5 August 2011) and Italy (5 August 2011): https://www.ft.com/content/1f4ed1fa-65ac-11e4-aba7-00144feabdc0. On the major policy shifts imposed on debtor states in areas beyond EU competence by 'Memoranda of Understanding': Andreas Fischer-Lescano, *Human Rights in Times of Austerity Policy: The EU Institutions and the Conclusion of Memoranda of Understanding* (Baden-Baden: Nomos, 2014), pp. 93ff.; Tuori and Tuori (*Eurozone Crisis*), pp. 99, 188. On public critiques of conditionality as undermining sovereignty, see e.g. European University Institute (EUI) (2014), 'Constitutional Change through Euro Crisis Law: France', http://eurocrisislaw.eui.eu, p. 33.

[17] See esp. EU Reg. 1174/2011, where penalties for ignoring Commission 'recommendations' are outlined as part of the 'Excessive Imbalances Procedure'.

These moves and others amounted in particular to a challenge to the procedures and norms of *democracy*, arguably the most basic at stake. At both national and EU level, institutional provisions designed to enable the public contestation of decision-making, including consultation periods and debates, were circumvented or compressed. The genesis of the Troika was in part an effort to sidestep the procedures of the European Parliament. Had the existing EU framework been relied on, its input would likely have been required in the evaluation and approval of policy changes. At the national level, basing treaties on non-EU law allowed governments to approve measures with a lower threshold of parliamentary support.[18] Some were introduced by decree, such that they would have legal force before a parliamentary vote, while others relied on unconventional methods such as votes-of-no-confidence that avoided the usual channels of accountability.[19]

Such pressures were compounded by the acceleration of decision-making. Leading executive figures would later be frank: 'at the height of the crisis, far-reaching decisions had often to be taken in a rush, sometimes overnight. In several cases, intergovernmental solutions were chosen to speed up decisions or overcome opposition.'[20] Financial transfers and major new policy regimes were often approved in a matter of hours—in the smaller national parliaments especially, but even the German Bundestag—and sometimes by a small committee.[21] When acting quickly is the condition of having influence, the virtues of debate, scrutiny and deliberation fall away. Making decisions in impromptu forums that lacked binding procedures was another way executives ensured others had little time to resist.

Not only the practicalities of contestation but its legitimacy too was compromised. Governments tended to present intergovernmental agreements as the first significant step in crisis management: contesting them in parliamentary settings could then be cast as undermining hard-won, fragile stability. Such moves encouraged complaints about procedure that further crowded out discussion of

[18] That the EFSF was under Luxembourg law meant some governments could sign it without consulting parliament: see e.g. EUI (2014), 'Constitutional Change through Euro Crisis Law: Austria', http://eurocrisislaw.eui.eu, p. 10.

[19] For the former, see e.g. Spain, where the 2010 Budget Law was introduced by royal decree, i.e. a law enacted by the executive with immediate effect and for times of 'urgent need': the Congress of Deputies had 30 days to ratify it, but could not make amendments (EUI (2014), 'Constitutional Change through Euro Crisis Law: Spain', http://eurocrisislaw.eui.eu, p. 22). Likewise in Italy the EFSF was passed by decree law (EUI (2014), 'Constitutional Change through Euro Crisis Law: Italy', http://eurocrisislaw.eui.eu, pp. 16, 33). For examples of no-confidence votes, see Romania (EUI (2014), 'Constitutional Change through Euro Crisis Law: Romania', http://eurocrisislaw.eui.eu, pp. 5, 19); cf. Beukers, de Witte and Kilpatrick (*Constitutional Change*), pp. 182ff., 204ff.

[20] European Commission (2015), 'Completing Europe's Economic and Monetary Union'.

[21] The Irish parliament was given just four hours to discuss the Fiscal Compact; the Bundestag found itself with less than 48 hours to approve the financial aid package for Greece agreed by Eurozone finance ministers and the IMF in late November 2012. France and Greece were two other countries where accelerated procedures were used to approve treaty amendments. See EUI (2014) reports on Ireland, Germany, France, and Greece respectively (available at http://eurocrisislaw.eui.eu).

substance.[22] And crucially, whether contestation was accelerated or even bypassed, the effect was to leave opposition parties looking complicit in decisions. Having little opportunity to challenge them, they became politically enmeshed with the decisions of executives.

While some of the measures enacted were of dubious legality, many were not obviously so. The reliance on ad hoc institutional arrangements unforeseen in treaty law, the application of strict conditions in return for loans, or the use of international agreements requiring a lower threshold of parliamentary support, were all moves that could be defended on legal grounds. Indeed, arguably a key structural feature of the transnational sphere is how much can done *without* legal suspensions and breaches. Due to factors we shall say more of, to do with weak levels of constitutional codification, it becomes possible to depart significantly from existing norms without always departing from existing law.

The Exceptionalist Imaginary

Central to the justification of such manoeuvres was the idea of *exceptional circumstances*. Unconventional steps were advocated on the grounds the situation was out of the ordinary, posing a threat to certain valued achievements. These included the Euro currency in particular, the EU more generally, and the socio-economic order it undergirded. The political response could depart from existing norms because the problems were unprecedented in nature and scale. One sees this clearly in the discourse of the time, with the recurrence of terms such as emergency, crisis, exceptional, and the like.[23]

This appeal to exceptional circumstances came with a number of related ideas. First to observe is how *necessity* becomes the principle of action.[24] Measures were often adopted not on the grounds they were intrinsically desirable, or the least bad of a series of options, but on the understanding they were the only course of action that could protect certain valued ends. They were actions of *last resort*, demanded by the situation rather than the outcome of choice and volition. As the Commission President observed with regard to one measure, 'the package we have agreed tonight is a comprehensive package that confirms that Europe will do what it takes

[22] A case study is: Catherine Moury and Adam Standring, '"Going Beyond the Troika": Power and Discourse in Portuguese Austerity Politics', *European Journal of Political Research* 56 (3) (2017). The Portuguese president urged that the country should be the first of all member-states to ratify stabilization measures so as to signal its seriousness: see EUI (2014), 'Constitutional Change through Euro Crisis Law: Portugal', http://eurocrisislaw.eui.eu, p. 33.

[23] 'Exceptional circumstances' were cited when the Council established the EFSF in May 2010, and with it the Troika arrangement. (See the press release of the Extraordinary Council meeting, Economic and Financial Affairs, Brussels, 9/10 May 2010, leading to the adoption of the EFSF.) 'Exceptional measures for exceptional times' were invoked by Commission President Barroso when responding to a further stabilization measure the following year (Barroso, 27 Oct. 2011, press statement).

[24] For discussion: Jonathan White, 'Emergency Europe' *Political Studies* 63 (2) (2015), p. 303.

to safeguard financial stability'.[25] Such statements affirmed the necessity of the measures, as well as de-contesting the ends they were to serve.

To view policies as necessary is to treat the ends attached as of general value. It is to see goals like financial stability and its related socio-economic effects as goals in which all are invested. In this sense one is dealing with claims—'emergency claims', as they have been called.[26] Prioritizing financial stability meant being willing to overlook who bore the sacrifices made—private investors or publics at large. Witnesses record that the ECB under Trichet convinced the German Chancellor and IMF in spring 2010 of the need to calm the markets irrespective of distributive considerations, adopting measures she explicitly characterized as of last resort.[27] Financial stability was thereby cast as a superordinate goal, and the interests of investors generalized as the common interest. Given such stakes are obscured by the talk of necessity, one may see this as a depoliticizing move. Even where one can credibly speak of 'necessary' policies, it remains an open question what policies these should be combined with. There are invariably choices to be made about what to do in addition; assertions of necessity tend to overlook this.

A further corollary is an emphasis on *speed*. If action is guided by necessity, there are no contending options to be weighed, and delay must be construed as loss of time. The Troika was a structure designed for speed, both in the sense of directing change in member-state law faster than conventional processes would allow, and ensuring the swift relay of information via reports to the relevant creditors and Council members. Informal forums such as the Eurogroup likewise promised rapid decisions in the face of deadlines. The importance of speed was used to put pressure on national governments, who in turn put pressure on national parliaments.[28] Where decrees were used, it was in the name of swift resolution.[29]

[25] Barroso press statement, 27 October 2011. See also the press release introducing the EFSF, stating it 'will stay in place as long as needed to safeguard financial stability' (press release of the Extraordinary Council meeting, Economic and Financial Affairs, 9/10 May 2010).

[26] Jennifer Rubenstein, 'Emergency Claims and Democratic Action', *Social Philosophy and Politics* 32 (1) (2015), pp. 101–26; cf. Giorgio Agamben, *State of Exception* (Chicago: Chicago University Press, 2005), pp. 1, 29; Brian Milstein, 'Thinking Politically about Crisis: A Pragmatist Perspective', *European Journal of Political Theory* 14 (2) (2015), pp.141–60.

[27] Ashoka Mody, *EuroTragedy: A Drama in Nine Acts* (Oxford: Oxford University Press, 2018), p.279; cf. Merkel's 'Ultima Ratio' justification, articulated i.a. on 5 May 2010: 'Die Grundlage für dieses Gesetz ist eine Ultima Ratio, also eine Notsituation. Die Notsituation besteht darin, dass Griechenland faktisch keinen Zugang zu den Finanzmärkten mehr hat. Daraus wären Auswirkungen auf die Stabilität des Euro insgesamt entstanden.' (https://archiv.bundesregierung.de/archiv-de/regierungserklaerung-von-bundeskanzlerin-merkel-zu-den-hilfen-fuer-griechenland-1122362).

[28] In his letter to the Irish government of 19 November 2010, ECB head Trichet urged that it seek a bailout from the Eurogroup and make 'rapid and decisive progress' in 'fiscal consolidation, structural reforms and financial sector restructuring': 'a swift response is needed before markets open next week'. The suggestion was that the ECB would withdraw funding from Irish banks if the Irish government failed to comply.

[29] EUI ('Constitutional Change through Euro Crisis Law: Italy'), p. 16: 'Decree-laws are legislative acts of a temporary nature having the force of law, adopted in "extraordinary cases of need and urgency" by the Government, pursuant to art. 77 of the Constitution of the Italian Republic.'

Sometimes, effectively invoking the need for speed depended on leaders going slow for a while. EU figures were often accused of hesitation, of a reluctance to embrace speed to the full—by representatives of the markets especially. Partly an effect of how politics generally may look slow beside the temporality of financial markets, arguably it was more than this. The more powerful decision-makers might deliberately hesitate at times so that the sense of urgency increased, making the threats to the socio-economic order appear greater—to the Eurozone in particular, to the lending regime more generally—and so the doctrine of last resort could more convincingly be deployed. Here was the logic of the *self-cancelling prophecy*: credibly raising the prospect of further chaos as a way to reassert control. German Chancellor Merkel was one figure suspected of dragging her feet in EU negotiations so as better to fend off opposition from the Bundestag, Bundesgericht, and wider German public.[30] The effect was to construct more clearly at key points a dichotomy of order and imminent disorder, giving dissenters strong reason to accept the government's position. Speed was selectively deployed, in other words, so that at critical moments the need for it could be impressed on others.

A diagnosis of exceptional circumstances is not just a way of characterizing the present. With it come certain ideas about the past and future. Regarding the former is the suggestion of a preceding period of normality, before the exceptional moment. Implied is the sudden onset of malaise, of problems newly arising rather than the slow fruition of long-term factors.[31] The singularity of the moment tends to be emphasized over regularities, precedents, and comparisons.

One way this outlook manifests itself is in explanatory accounts. Two kinds recurred in the discourse of the Euro crisis. On the one hand, the origins of misfortune were commonly described like a natural disaster, arising from 'exceptional circumstances beyond Member-States' control'.[32] Such formulations cultivated the idea of a crisis striking from nowhere, and discouraged or postponed the analysis of causes. Their appeal no doubt lay partly in the capacity to remove responsibility from the problem definition. In transnational politics especially, fatalistic narratives may be a valuable resource, since negotiations may run more smoothly if blame can be set aside.

On the other hand, and pointing in a different direction, the origins of malaise were presented as the consequence of *mistakes*, as visible in condemnations of

[30] Martin Sandbu, *Europe's Orphan: The Future of the Euro and the Politics of Debt* (Princeton: Princeton University Press, 2015), p. 59. See also J. H. H. Weiler, 'Integration through Fear', *European Journal of International Law*, 23 (1) (2012), pp. 1–5.

[31] Cf. Craig Calhoun, 'A World of Emergencies: Fear, Intervention and the Limits of Cosmopolitan Order', *Canadian Review of Sociology* 41 (4) (2004), pp. 373–95, who speaks of an 'emergency imaginary' emphasizing 'apparent unpredictability, abnormality and brevity'.

[32] See e.g. the announcement of the EFSF. The phrasing was partly shaped by the technical concern to square the mechanism's existence with 122.2 of the Lisbon Treaty.

state corruption and profligacy.[33] Treating the source of crisis as lying in the transgressions of defined actors, and the solution as in Troika-style measures to get them to change their ways, was one way of retrospectively normalizing the pre-crisis period.[34] As with natural-disaster accounts, such a perspective discourages the thought that the problems are deep-seated, and that those taking the lead in responding to the situation might in some way be co-responsible for its production. For those engaged in norm-defying actions, excluding such possibilities will be a natural aim, since the moral licence for extreme measures might seem weaker if the problems at hand were foreseeable or of their own making.[35]

Were more far-reaching explanations not also present? It was commonly observed that the design of the Eurozone was imperfect, notably as regards the separation of monetary and economic policy, suggesting the recognition of endemic factors. Yet two things should be noted. First, such diagnoses often emphasized not that the pre-existing order was faulty but 'incomplete'. The EU needed to establish 'the necessary reforms to *complement the existing framework*', as the EFSF announcement would explain.[36] Such formulations implied the need for additions rather than subtractions, and thereby defended the essentials of the preceding order. Second, though more fundamental diagnoses did emerge, focused not just on remedial measures but preventive ones to ward off the next crisis, these tended to arrive after the initial course had been set. They worked within the parameters set by the first initiatives.

If 'exceptional measures for exceptional times' suggests a *prior* period of normality, it also suggests a period of normality-to-come. The suggestion is that unconventional rule will be temporary.[37] Edging towards this 'posterior normal' might take different forms. It might involve discontinuing certain measures associated with exceptional times, and rejecting the notion they set precedents. Equally it might involve building on them, normalizing them through retrospective ratification or incorporation into a new framework. Change might be viewed as necessary to prevent restoring the conditions for crisis.[38] One way or another,

[33] See ECB Head Draghi, press conference, 6 September 2012, Frankfurt: 'We should not forget why countries have found themselves in a bad equilibrium to start with. And this is because of policy mistakes...'

[34] Jonathan White, 'Left and Right in the Economic Crisis', *Journal of Political Ideologies* 18 (2) (2013), pp.150–70.

[35] On moral intuitions concerning the relation between responsibility and exceptionalism: Tom Sorell, *Emergencies and Politics: A Sober Hobbesian Approach* (Cambridge: Cambridge University Press, 2013). One of the felicitous aspects of emergency situations for those in power is that the search for deeper causes may seem ill-advised, even irresponsible, when time is at a premium: see Chapter 9.

[36] EFSF press release, as above; my ital.

[37] See e.g. Commission President Juncker, policy guidelines for the new Commission, 2014, 'New Start': http://www.eesc.europa.eu/resources/docs/jean-claude-juncker—political-guidelines.pdf.

[38] Cf. Jon Elster, 'Comments on the Paper by Ferejohn and Pasquino', *I.CON* 2 (2) (2004), pp. 240–3; see also the discussion of emergency rule later in this chapter. Cf. Sorell (*Emergencies and Politics*), p. 153, on 'the old normal, the new normal and the abnormal'.

termination of the irregular is implied. This suggestion of the *temporary* character of the circumstances at hand and the form of rule they demanded would have important power-political effects.

Taken collectively, these ideas of exceptional circumstances, the demands of necessity, the importance of speed, and an anterior and posterior normality, amount to a collective imaginary central to the governing rationality of the time. Such an outlook is composed of contestable claims about the way things are and can be expected to go. Clearly, the problems at hand need not be fictitious. As later chapters examine, there were material drivers of this outlook and the practices it sought to rationalize. The logic of exceptionalism suggested itself because of objective features of the situation in question, but to interpret circumstances this way was a move not objectively determined.

The Hour of the Executive

Exceptionalism amounts to a notion that political interventions can make a difference, that decisions addressed to extreme situations can be effective, so long as they are fast and adaptive. It is a call to action, albeit in the weak sense of executing the demands of necessity.[39] In such an ideational context, it is predictably executive agents who come to the fore. Conventional wisdom suggests that when exceptional circumstances are declared and the constraints of procedure relaxed, it is these fast-moving, centralized, and unified agents who rise to the occasion.[40] Heightened executive activity ensues, and it proceeds on unpredictable terms. Governments adopting the emergency mode may feel entitled to do things previously unadvertised, exceeding or breaking for instance their manifesto commitments. In circumstances understood to be unexpected and even unprecedented, officials can do what was previously off limits.

As I want to argue, these patterns were fully present in the case at hand, yet executive activism here took on a distinctive form due to the structure of the transnational sphere. The many institutions implicated in making and implementing decisions—national, sub-national and supranational, along with multilateral and bilateral gatherings—meant power was *dispersed across* many sites. It was also *concentrated within* institutions on individuals and sub-groups, acting outside their formal roles and detached from their organization's standard procedures, tending in the limit case to what I call 'de-institutionalization'. This was executive power as the power of networks.

[39] Ben Anderson, 'Emergency Futures: Exception, Urgency, Interval, Hope', *Sociological Review* 65 (3) (2017); also Chapter 6, this volume.

[40] William Scheuerman, *Liberal Democracy and the Social Acceleration of Time* (Baltimore: Johns Hopkins University Press, 2004); Posner and Vermeule (*Executive Unbound*).

We have noted the degree to which supranational executive agents such as the Troika took on new powers of oversight and implementation, and how ad hoc intergovernmental forums such as the Eurogroup became central. The demand for speed favoured the creation of arrangements like these that were unencumbered by procedural constraints. But extensions of executive power at the European level were coupled with, and depended on, similar extensions into member-states. Debates about whether supranational or intergovernmental authority was most strengthened in this period risk overlooking how both tendencies interplayed with domestic ones.[41] Empowerment of the Troika at the supranational level went hand-in-hand with the empowerment of national executives—not just those with the strongest voice on the Council, but those in smaller states on the receiving end of Troika policies. The pressures for speed encouraged the build-up of domestic interlocutors with whom Troika representatives could interact—national governments generally, and finance ministries especially.[42] These agents in turn could use external demands as reason to press ahead with measures against domestic resistance, as the governments of Monti in Italy and Papademos in Greece would show.[43] Supranational, transnational and national executive forms interlocked.

An implication of this dispersal of power was that each institutional agent could present itself as compelled to act by others, as a pawn in a larger game. The fact of Troika intervention could be used by governments to introduce measures additional to those demanded,[44] while the *prospect* of its intervention could be used to advocate austerity measures in advance of any formal requirement.[45] The dynamics of surveillance and anticipatory compliance were both present and strategically invoked. Likewise Troika figures could claim to be constrained by the agenda of member-states, or indeed peers in the Troika, each formally with its own procedures to respect. The dispersal of executive power, partly structural and partly the result of deliberate efforts to multilateralize, served to

[41] Cf. Deidre Curtin, *Executive Power of the European Union. Law, Practices and the Living Constitution* (Oxford: Oxford University Press, 2009); Beukers, de Witte and Kilpatrick (*Constitutional Change*), pp. 108ff.

[42] See e.g. Moury and Standring ('Going beyond the Troika'); also EUI (2014), 'Constitutional Change through Euro Crisis Law: Greece', http://eurocrisislaw.eui.eu, p. 20; cf. EUI (2014), 'Romania', pp. 15ff. on the strengthening of the Romanian government by the 2010 Fiscal Responsibility Law, pushed by the Troika.

[43] This option could also be waived: sometimes crisis measures were presented, for reasons of national dignity (or to avoid constraints on the transfer of authority to international institutions), as *national* necessities rather than Troika requirements: see e.g. EUI (2014), 'France', p. 15; EUI (2014), 'Greece', pp. 26, 97–8, 107). The pattern was the *selective* externalizing and internalizing of responsibility.

[44] See e.g. the Portuguese government trying to privatize further than its Troika MoU required (EUI Report on Portugal, pp. 47–8).

[45] See EUI (2014), 'Constitutional Change through Euro Crisis Law: Slovenia', http://eurocrisislaw.eui.eu (p. 41) on the introduction of the Six-Pack.

anonymize it, raising problems for those who would challenge it and those who would seek later to reverse its decisions.

The dispersal of power was paralleled by its concentration within institutions at the apex. One saw this both in supranational and intergovernmental contexts. If one reads official evaluations of the Troika arrangements—of the Commission by the Court of Auditors, of the IMF by its own assessors[46]—the pattern is not so much of institutions acting in regularized fashion, as 'the Commission', 'the IMF' and so on, as of individuals and groups taking decisions independently, improvising their methods and excluding some in the process.[47] Likewise, state representations in the Council were, in the most accelerated moments of decision-making, reduced to small coteries acting in relative isolation. Initiatives aimed at setting up new treaties were taken by (prime) ministers and advisors largely detached from the influence of their wider civil service and negotiating teams, often at extreme hours of the day. The fact that some of these were ad hoc forums, convened specifically to achieve fast decisions and without the involvement of supporting negotiators, contributed to the personalization effect.

This individualization of power within institutions was matched by the development of close *cross-cutting* ties, again at the apex especially. Informal collaboration between those attached to distinct organizations was common. The networks associated with the Troika were one manifestation. In-house reflections on the Troika stress informal ties as the precondition of its functioning.[48] Ad hoc gatherings such as the 'Frankfurt group', featuring the heads of the IMF, ECB, Commission, and Eurogroup, as well as the French President and German Chancellor, likewise spoke to the importance of cross-institutional ties and close working relationships between individuals at the interstices of networks.[49]

[46] European Court of Auditors, 'Financial Assistance Provided to Countries in Difficulties', Special Report 18 (2015); IMF Report 'The IMF and the Crises in Greece, Ireland and Portugal: An Evaluation by the Independent Evaluation Office', 8 July 2016.

[47] An IMF report criticizes decisions taken informally within the Fund by a small group, leaving some board members feeling excluded (Independent Evaluation Office of the IMF ('IMF and the Crises'), pp. viii, 18, 47). Such practices are the radicalization of pre-existing trends towards 'presidentialization' in many of the institutions: see Stefan Becker, Michael W. Bauer, Sara Connolly, and Hussein Kassim, 'The Commission: Boxed in and Constrained, but Still an Engine of Integration', *West European Politics* 39 (5) (2016), pp. 1011–31.

[48] See e.g. the 2015 Court of Auditors' report on the Commission ('Financial assistance provided to countries in difficulties') on the informality of cross-institutional ties (pp. 47, 75); likewise IMF ('The IMF and the Crises . . . '), p. 42; cf. G. Russell Kincaid, 'The IMF's Role in the Euro Area Crisis: What are the Lessons from the IMF's Participation in the Troika?', *Independent Evaluation Office of the IMF*, Background paper BP/16-02/06 (2016), p. 31; Henning (*Tangled Governance*), p. 34.

[49] On the close personal ties between Juncker (of the Commission) and Schulz (of the EP), to the irritation of members of both institutions: Desmond Dinan, 'Governance and Institutions: A More Political Commission', *Journal of Common Market Studies* 54 (2016), Annual Review, p. 102; cf. Henning (*Tangled Governance*), p. 232. On the Frankfurt Group, Antoine Vauchez, *Democratizing Europe* (Basingstoke: Palgrave Macmillan, 2016), p. 17. On some critical decisions in spring 2010 being taken informally in small groups: Tony Barber (2010), 'The Euro: Dinner on the Edge of the Abyss', FT.com, 10 October.

Indeed, rather than think in terms of the empowerment of an institution, it may be useful to see this as *de*-institutionalization.[50] An ontology of institutions depends on several assumptions, including that there exists a fairly settled set of practices giving shape to an institution, and that the ties of interaction between those *within* each are denser than those across. EU scholarship of both the intergovernmental and neofunctionalist variety tends to rely on this ontology. Yet these were exactly the assumptions challenged. Institutional resources did not melt away, but were combined in unconventional ways spanning multiple sites of decision. Ideas of principal–agent relations, and more generally of representation, become hard to sustain in such contexts. It is exactly the familiar units of analysis—the procedure and the institution—that cease to be reliable guides.

These cross-cutting arrangements were something that could strengthen the hand of executive power. They allowed agents to draw on a wider range of resources, to take on different roles at different times, and to use the material and symbolic resources of one institution to bolster another. They spared them critical scrutiny, since formal roles could then offer little insight into where responsibility for decisions lay.[51] The capacity to distinguish institutions and offices from the persons that hold them was weakened. Individuals could disclaim involvement or point to others as the deciding influence. At the same time, this distributed structure raised enduring challenges for the coherence of executive action, and posed questions for the ideals of technocracy on which decisions had hitherto relied.

Executive power took network form. One of the artefacts of this period was the phenomenon of the 'Four Presidents', later the 'Five Presidents', in whose name several blueprints for the future were released, presented as reflecting their 'personal deliberations and discussions'.[52] Here once more was an agenda-setting group that was centralized and informal, and with a shape-shifting quality. Like the Troika, it was very much a creature of its times, illustrating what the hour of

[50] This promising term has been used before in EU studies, but with the focus on lower-level networks of influence around the Commission: Arthur Benz, 'Compounded Representation in EU Multi-level Governance', in Beate Kohler-Koch (ed.), *Linking EU and National Governance* (Oxford: Oxford University Press, 2003), p. 101.

[51] Cf. Varoufakis (*Adults in the Room*, p. 308) on the 'runaround', whereby critical agents were systematically inhibited from identifying appropriate interlocutors. One is reminded of the dangers highlighted by the American Federalists about the diffusion of executive power: as Hamilton wrote in *Federalist Papers* 70 (pp. 348–9), 'the multiplication of the executive adds to the difficulty of detection [of responsibility] ... It often becomes impossible, amidst mutual accusations, to determine on whom the blame or the punishment or a pernicious measure, or series of pernicious measures, ought really to fall. It is shifted from one to another with so much dexterity, and under such plausible appearances, that the public opinion is left in suspense about the real author.'

[52] Five Presidents' blueprint for the future of the Eurozone, 1 July 2015, http://europa.eu/rapid/press-release_IP-15-5240_en.htm. The original 'Four Presidents' (responsible for the 5 December 2012 report 'Towards a Genuine Economic and Monetary Union') were van Rompuy (Council), Barroso (Commission), Juncker (Eurogroup), and Draghi (ECB). The 'Five Presidents' also featured Draghi (ECB) and Juncker (now as head of Commission), together with Tusk (Council), Dijsselbloem (Eurogroup) and Schulz (EP).

the executive in this context might mean. Whereas the familiar image of executive empowerment sees power concentrated in the hands of the President, here one saw it concentrated in the hands of the Five Presidents.

Conceptual Foundations: Emergency Rule in the Transnational Setting

There is a structural unity, I suggest, to the tendencies described. A politics that escapes norms finds its rationale in exceptionalism—in the notion that exceptional circumstances warrant unconventional moves to maintain order. Both combine naturally with the foregrounding of executive agents, as those best placed to respond quickly, take decisions and enforce them. This unity is well captured by a long-standing concept, adaptable to the transnational context. What the EU was witness to was a form of *emergency rule*, albeit not quite as we know it.[53]

One sees this better if we put our preceding observations in a larger theoretical context, starting with the notion of a politics *escaping norms*. To step back in time for a moment: in the political thought of the ancient Roman republic, one finds clear anticipation of the prospect that rulers might find themselves in situations where tempted to break with the norms of good government. The famous institution of *dictatorship*—understood as an exceptional mode of rule without pejorative connotations—was an effort to regularize the mechanisms by which this would happen. At least in its idealized form, of the kind later celebrated by Machiavelli, this was a system based on a clear separation between the normal order and the period of dictatorship, when existing laws could be suspended and new measures improvised (though not with the status of law).[54] Here was an idea

[53] In the book I use 'emergency politics' and 'emergency rule' largely interchangeably, though the former is preferred when including actions beyond executive institutions. Taking a similar angle on the Euro crisis, though optimistically regarding it as an opportunity for the EU's positive transformation: Ernst-Wolfgang Böckenförde (2010), 'Kennt die europäische Not kein Gebot?', *Neue Zürcher Zeitung*, 21 June (https://www.nzz.ch/kennt_die_europaeische_not_kein_gebot-1.6182412). Closer to my perspective is Florian Rödl (2012), 'EU im Notstandsmodus', *Blätter für Deutsche und Internationale Politik* 5 (5) (https://www.blaetter.de/archiv/jahrgaenge/2012/mai/eu-im-notstandsmodus), who speaks of an 'emergency mentality' in the EU: 'Regierungen und Organe der Union agieren im Rahmen einer ungeschriebenen Notstandsverfassung. Dies muss, auch wenn natürlich jedem die Bewältigung der europäischen Finanzkrise in der Sache sehr am Herzen liegt, aus rechtsstaatlichem und demokratischem Blickwinkel zutiefst beunruhigen.' See also Joseph Vogl, *The Ascendancy of Finance* (Cambridge: Polity, 2016), pp. 5ff.; Giorgio Agamben, 'The Endless Crisis as an Instrument of Power', 4 June 2013: https://www.versobooks.com/blogs/1318-the-endless-crisis-as-an-instrument-of-power-in-conversation-with-giorgio-agamben; Jan Suntrup, 'From Emergency Politics to Authoritarian Constitutionalism? The Legal and Political Costs of EU Financial Crisis Management', *German Law Journal* 19 (2) (2018).

[54] Niccolò Machiavelli, *Discourses on Livy*, trans. Julia Conaway Bondanella and Peter Bondanella (Oxford: Oxford University Press, 2008/1531). The model was perhaps rarely realized even in the early Roman Republic; in later years it was markedly not, as the institution of dictatorship crossed into permanent autocracy. For attempts to correct later mythologizing about the Roman institution: Andrew Lincott, *The Constitution of the Roman Republic* (Oxford: Oxford University Press, 2003);

of unconventional politics centred on formally suspending the legal order, in a circumscribed fashion, and for the sake of maintaining that order.

Later thinkers, notably Carl Schmitt, would cast doubt on the extent to which emergency rule could, and should, be confined in this way. The suggestion was that efforts to codify the interruption of law were mistaken, since the conditions that might warrant such interruption were those that resisted constitutional anticipation. Schmitt's writings suggest a more open-ended kind of emergency rule, with the Roman model reinterpreted as just one, largely antiquated, form.[55] Others, such as Carl Friedrich, saw the challenge as how to update the principles of the Roman model for a modern industrial society.[56] Generally such perspectives retained a focus on the formal suspension of (aspects of) the legal order, enacted by triggering an emergency clause in the polity's constitution.

In the EU's handling of the Euro crisis one saw a more informal kind of emergency rule. There was no legal clause to trigger, and no formal suspension of the existing legal order.[57] While rule by decree was seen in some national contexts, offering a conventional marker of emergency rule, elsewhere it was absent. And though the measures taken at the EU level were sometimes at odds with existing EU and national law, at other times they simply took advantage of the possibilities the law left open. In contrast to many of the classical theories, here the escape from norms was the escape from political norms, either without legal expression or—to the extent they appeared in the EU treaties—expressed in quite general terms.[58]

To be sure, nation-states have also seen periods of emergency rule that are informal and undeclared, and a tendency to work with and around the law.[59]

Marc De Wilde, 'The Dictator's Trust: Regulating and Constraining Emergency Powers in the Roman Republic', *History of Political Thought* 33 (4) (2012), pp. 555–77; Wilfried Nippel, 'Saving the Constitution: the European Discourse on Dictatorship', in Janet Coleman and Paschalis M. Kitromilides (eds.) *In the Footsteps of Herodotus: Towards European Political Thought* (Florence: Leo S. Olschki, 2012), pp. 29–50; Marco Geuna, 'Machiavelli and the Problem of Dictatorship', *Ratio Juris* 28 (2) (2015), pp. 226–41; Andreas Kalyvas, 'The Tyranny of Dictatorship: When the Greek Tyrant Met the Roman Dictator', *Political Theory* 35 (2007).

[55] This unboundedness comes through in the discussion of sovereignty in *Political Theology*; reflections on the Roman model appear as an account of 'commissarial dictatorship' in *Dictatorship* (see n. 59).

[56] Carl Friedrich, *Constitutional Government and Democracy: Theory and Practice in Europe and America* (Waltham, Mass: Blaisdell, 1968), p. 581.

[57] See Joerges in Damian Chalmers, Markus Jachtenfuchs and Christian Joerges (eds), *The End of the Eurocrats' Dream: Adjusting to European Diversity* (Cambridge: Cambridge University Press, 2016); Beukers, de Witte, and Kilpatrick (*Constitutional Change*), chs 6–9.

[58] Likewise connecting emergency logic to a departure from norms: Sorrell (*Emergencies and Politics*, p. 1).

[59] From a large literature: Oren Gross and Fionnuala Ní Aoláin, *Law in Times of Crisis: Emergency Powers in Theory and Practice* (Cambridge: Cambridge University Press, 2006); John Ferejohn and Pasquale Pasquino, 'The Law of Exception: A Typology of Emergency Powers,' *International Journal of Constitutional Law* 2 (2004); Nomi Claire Lazar, *States of Emergency in Liberal Democracies* (Cambridge: Cambridge University Press, 2009). Friedrich (*Constitutional Government*) emphasized

Conceptually, Schmitt observed that dictatorship might entail departures from standards that were weakly codified: 'what is seen as the norm can positively be defined either by an existing constitution or by a political ideal'; departures could be from 'an existing or imagined constitution.'[60] But these informal features become more pronounced in the transnational setting, given its weak constitutionalization, the use of constructive ambiguity in treaty-making, and the ever-present possibility of stepping outside existing arrangements and reverting to international relations.

There is a risk that, by speaking of 'informal' emergency rule, one strays too far from the classical paradigm. Arguably there was a good reason why positive law was typically at the centre of these theories: it was against this backdrop that with some precision one could discern departures from the political order's norms. How can one be confident in identifying transgressions and evasions if these are not clearly signalled in legal terms? Why not see periods of unconventional decision-making as simply an extension of the EU's standard ways, in no sense to be viewed as special? 'Any exception,' wrote Schmitt, 'from a condition conceived of as normal can be called a dictatorship; and so this word can designate an exception from democracy; from constitutionally guaranteed liberties; from the separation of powers...'[61] But how can we be sure such things were still 'conceived of as normal' in Europe by the first decade of the twenty-first century?

My assumption is that one finds confirmation of the departures described from the level of critical response they provoked from interested observers. Though norms pose a challenge of identification and may be subject to competing interpretations, one thing that attests to their transgression is the condemnation arising when agents depart from them. The actions described were all the subject of controversy amongst parliamentarians, media observers, scholars, and others.[62] This capacity to provoke a measure of outrage reminds one that, however imperfectly such standards were upheld in the pre-crisis EU order, what happened in this period marked a significant breakdown. This was not simply business as usual. But nor need one assume there were not continuities. Here, as typically, emergency rule marks the radicalization of existing tendencies.

the discrepancy between theories of emergency rule and its practice in modern states: as he observed (p. 571), in many respects 'the traditional view is too legalistic'.

[60] Schmitt, *Dictatorship*, pp. xli, 128. These distinctions are tied in his thought to another, between restorative ('commissarial') and transformative ('sovereign') dictatorship, which I do not adopt given the challenges of empiricizing it (of which Schmitt was arguably aware: William Scheuerman, 'States of Emergency', in Jens Meierhenrich and Oliver Simons (eds), *Oxford Handbook of Carl Schmitt* (Oxford: Oxford University Press, 2017); Duncan Kelly, 'Carl Schmitt's Political Theory of Dictatorship', in Jens Meierhenrich and Oliver Simons (eds), *Oxford Handbook of Carl Schmitt* (Oxford: Oxford University Press, 2017)).

[61] Schmitt, *Dictatorship*, p. 128.

[62] See again the country reports available at eurocrisislaw.eui.eu.

In the prevalence of the *exceptionalist imaginary*, again one sees a familiar feature of emergency rule adjusted to the EU setting. The appeal to exceptional circumstances was part of the classical rationale for dictatorship. For the Romans, Machiavelli, Rousseau, Schmitt, and others, these circumstances tended to be conceived as war or rebellion. Emergency rule has had a geopolitical focus, and retains this in contemporary scholarship looking at the dynamics of the 'war on terror' and instances of humanitarian crisis. But the idea that other domains such as the economic might fall within the purview of emergency rule is not new. As analyses of emergency politics in Weimar Germany and the Depression-era US make clear, an 'economic state of emergency' was a template already pioneered in the national setting.[63]

Yet its adoption in the EU is a little different. In the state and proto-state context, the addressee of emergency claims is generally assumed to be a domestic population, urged to accept intrusions on its rights or values. This remains a central focus in the transnational context: a domestic public is often one of the addressees, encouraged to grasp the necessity of the measures imposed. But equally, given the dispersal of power across multiple institutions, the addressee may involve other decision-makers, thereby pressured to coordinate and act.[64] The identification of exceptional circumstances can be scaled up from the national to a pan-European level, such that an apparently local problem is elevated to a common threat (and later re-localized once more). The emergency can be projected outwards.[65] Given the extent of interdependence, the addressee may also include those in more than one state (socio-economic groups, or general populations) invited to observe that major initiatives are being taken and to adjust their behaviour. There is an aspect of public performance to this outlook, accentuated in the transnational context.

It should be emphasized that these discursive aspects of emergency rule are politically consequential. As scholars of 'securitization' have taught us,[66] it is by deploying such interpretative motifs to rationalize their decisions that agents can influence how others respond—whether they seek to contest these decisions, ignore them, or adapt. To emphasize the urgency, necessity, and temporariness of initiatives is to question the reasonability and legitimacy of attempts to oppose them, as well as to suggest they need not be argued on their merits. Such

[63] On Weimar and the US: Clinton Rossiter, *Constitutional Dictatorship: Crisis Government in the Modern Democracies* (Princeton: Princeton University Press, 1948); on the 'economic state of emergency': William Scheuerman, 'The Economic State of Emergency', *Cardozo Law Review* 21 (5–6) (1999).

[64] See e.g. the Trichet letters.

[65] Barber (2010) reports a summit of leaders, 7 May 2010: 'One participant recalls: "Trichet said: 'This isn't only a problem for one country. It's several countries. It's Europe. It's global. It's a situation that is deteriorating with extreme rapidity and intensity."'

[66] Barry Buzan, Ole Wæver, and Jaap de Wilde, *Security: A New Framework for Analysis* (Boulder CO: Lynne Rienner, 1998).

statements help shape the political culture of the moment. Particularly if we are interested in questions of power and democracy, these rationalizations are key. They are constitutive aspects of emergency rule, not just rhetorical colour.

The third element of the discussion—the *hour of the executive*—is again characteristic of emergency rule, given distinctive inflection in the transnational context. As Clinton Rossiter once wrote, 'it is never so apparent as in time of crisis that the executive is the aboriginal power of government'.[67] He, like others, assumed the executive to be a unitary agent. On Machiavelli's reading, 'the Romans, among the other customary remedies taken during times of urgent danger, decided to create a dictator, that is, to give power to a man who might make decisions without any consultation whatsoever and who, without any right of appeal, might execute his decisions'.[68] Power was concentrated in the singular official.[69] In his work on the figure of the sovereign who would emerge in exceptional times, Schmitt tends to be reticent on this agent's identity (often offering just the pronoun 'he'), but it seems clear that he likewise views the figure as indivisible.[70]

Conversely, the EU case shows that executive empowerment in the transnational setting is hard to localize on a single individual or institution. In place of the dictator on the classical model, embodied in a unitary agent, one saw a precarious process of coordination across different institutions—the Troika again being emblematic. The relevant collective spanned a wide institutional field, from leading figures of the ECB to the Commission, the Council to member-states. What authority they had derived precisely from their *interactions*, as the statements of some shored up the actions of others, granting them credibility or withholding it. Emergency rule was co-produced by the many, as EU figures relied on executive power extending into the member-states, while agents of the latter selectively externalized responsibility back to intergovernmental and supranational institutions. While the behaviour of certain institutions can productively be analysed using Schmittian ideas of dictatorship, the EU field as a whole resists such analysis. This was emergency rule without a defined institutional sovereign—a dispersed emergency regime rather than a clearly authored state of exception.[71]

As we shall see, this matters on several counts. The dispersal of executive power caused heightened problems for its democratic contestation, as well as for the kind

[67] Rossiter (*Constitutional Dictatorship*), p. 13.

[68] Machiavelli (*Discourses*), p. 91 (i.33). Cf. Nippel ('Saving the Constitution'), p. 31.

[69] In the ancient Roman case as read by Machiavelli, the *declaration* of exceptional circumstances was to be in the hands of a different agent (a public council).

[70] *Political Theology*, p. 10 suggests some possibilities—God, the emperor, the prince, or 'the people, meaning those who identify themselves directly with the people': Carl Schmitt, *Political Theology: Four Chapters on the Concept of Sovereignty* (Chicago: Chicago University Press, 2005).

[71] See White ('Emergency Europe'); William Scheuerman, '*Die Globalisierung von Carl Schmitt?*', *KJ Kritische Justiz* 50 (2017a), p. 35. Challenging Schmitt's emphasis on sovereign decision and highlighting more diffuse acts of discretion in emergency's name: Bonnie Honig, *Emergency Politics: Paradox, Law, Democracy* (Princeton: Princeton University Press, 2009), pp. xviii, 67.

of technocratic logic that has traditionally guided transnational decision-making. Because the legacy of emergency measures in these years has been to give long-term discretionary oversight to the Commission, Council, and ECB,[72] these features are now embedded in the system, with important constitutional implications. They are also one of the reference-points for public disaffection.

What was witnessed in the EU in the 2010s was a form of emergency rule, albeit one that challenges assumptions carried over from the state context. That the measures taken seemed to prioritize reinforcing the socio-economic order more than the EU's political form is consistent with this. It has long been an observation (and concern) that emergency exceptionalism can be indifferent to constitutional aims and oriented to 'order' in a more general sense.[73] Likewise, that the EU evolved significantly in these years is very much in line with the emergency perspective. While, following the ancient Roman accounts, emergency rule has often been defended on the grounds it is merely *restorative*, the reality has tended to look different. To restore the preceding order would be to restore the conditions that required exceptional measures.[74] Not just the circumstances at hand, but a concern to ward off their future recurrence, has tended to inform the measures taken.

For the sake of reinforcing the foundations of the existing socio-economic order, the EU's institutional order has seen far-reaching change. Indeed, precisely because of this legacy of institutional redesign, the politics of emergency has lasting significance. In the transnational context especially, there is an aspect of polity-building to it.[75] Produced by the many and with effects that are difficult to undo, this is something other than mere interruption, other than a sovereign state of exception.

Theoretical Contribution

As the above may suggest, there is a contribution to be made to theories of emergency rule by making them receptive to the dynamics of the transnational setting, on which existing work remains thin. Emergency rule has tended to be tied

[72] Agustín José Menéndez, 'The Crisis of Law and the European Crises: From the Social and Democratic *Rechtsstaat* to the Consolidating State of (Pseudo-)Technocratic Governance', *Journal of Law & Society* 44 (1) (2017), pp. 56–78.

[73] On this basic distinction concerning the ends to which emergency rule is geared: Friedrich (*Constitutional Government*), p. 560. On the socio-economic concerns shaping Europe's crisis response in the 2010s: Claus Offe, *Europe Entrapped* (Cambridge: Polity, 2014).

[74] See also Friedrich (*Constitutional Government*), pp. 560, 565; cf. Ulrich K. Preuß, 'Die Krise der Europäischen Union als Ausnahmezustand?', *KJ Kritische Justiz* 50 (2017), pp. 51–67. On the modern idea of dictatorship as transformative: Reinhart Koselleck, *Futures Past: On the Semantics of Historical Time* (New York: Columbia University Press, 2004), pp. 250–1.

[75] On 'emergency state-building': Tyler M. Curley, 'Models of Emergency Statebuilding in the United States', *Perspectives on Politics* 13 (3) (2015).

conceptually to sovereign power, and the time seems ripe to decouple them.[76] A closer look at the EU should give a better feel for the assumptions made by theories of emergency rule as developed for the national sphere. It should indicate how such assumptions need to be refined when looking beyond this sphere, but also how they may need to be revised if one is properly to understand the contemporary state. Insofar as the transnational increasingly permeates domestic affairs, the EU is a productive context for developing expectations of wider application.

What, on the other hand, can this perspective bring to the field of EU studies? In what sense is it not just feasible but advantageous to approach the EU in these terms? Notions of integration-through-crisis have been a staple of thinking about the EU's development since the days of Jean Monnet[77]—so what is to be gained from these ostensibly similar ideas?

One reason to take this view is to foreground the normative stakes. There is a long-standing tendency in EU scholarship to narrate the Union's history as one of successful adaptation to functional demands. Moments of crisis in EU history are commonly viewed as opportunities for overcoming deficits—of capacity, of will, of public awareness. Accounts of EU executive consolidation in the face of crisis 'events' are a recent extension of this view, showing a sophisticated understanding of statecraft but a tendency to validate the outcome.[78] As Christian Kreuder-Sonnen puts it, such integration-through-crisis perspectives tend towards ex-post rationalization: 'the crisis decisions thus not only appear as geared towards the right overall aim but also as logical and overdue'.[79] They smooth the edges of things, suggesting reasonable agents could do no other. Beholden to the logic of unintended consequences, verging on the cunning of reason, such views lose sight of the agendas and power relations in play, and of the unequal outcomes

[76] See also Calhoun ('World of Emergencies') and EU scholarship as below. For like-minded reappraisals of emergency politics today, within and beyond the state: Honig (*Emergency Politics*), esp. pp. 66ff.; Peter Adey, Ben Anderson, and Stephen Graham (2015), 'Introduction. Governing Emergencies: Beyond Exceptionality', *Theory, Culture & Society* 32 (2) 3–17; Sven Opitz and Ute Tellmann (2015), 'Future Emergencies: Temporal Politics in Law and Economy', *Theory, Culture and Society* 32 (2); Matthias Lemke, 'What Does State of Exception Mean? A Definitional and Analytical Approach', *Zeitschrift für Politikwissenschaft* 3 (online early).

[77] Monnet, *Memoirs*, p. 417. On crisis and European integration: Ludger Kühnhardt (ed.) *Crises in European Integration: Challenges and Responses, 1945–2005* (Oxford: Berghahn, 2009); Mai'a Davis Cross, 2017), *The Politics of Crisis in Europe* (Cambridge: Cambridge University Press, 2017). Two promising additions, although still argued largely in terms of elite adaptation to functional necessity, are: Erik Jones, R. Daniel Kelemen, and Sophie Meunier, 'Failing Forward? The Euro Crisis and the Incomplete Nature of European Integration', 49 (7) (2016), pp. 1010–34; and Lisbet Hooghe and Gary Marks (2018). 'Cleavage Theory Meets Europe's Crises: Lipset, Rokkan, and the Transnational Cleavage', *Journal of European Public Policy* 25 (1), pp. 109–35.

[78] Van Middelaar (*Alarums and Excursions*), p. viii, on recent improvisations as 'a sign of Europe's political maturity'; cf. p. 59 on 'how success was achieved in the end' in the Euro crisis.

[79] Christian Kreuder-Sonnen, 'Beyond Integration Theory: The (Anti-)Constitutional Dimension of European Crisis Governance', *Journal of Common Market Studies* 54 (6) (2016), pp. 1350–66. On the pitfalls of integration-through-crisis: Waltraud Schelkle, *The Political Economy of Monetary Stability: Understanding the Euro Experiment* (Oxford: Oxford University Press, 2017).

produced. Thinking instead in terms of emergency politics retrains one's focus, highlighting the ideas involved, the means adopted, and with what implications and lasting effects.

As others have noted, different political responses to economic malaise were available in this period: things did not have to be approached as they were. Instead of treating the prospect of defaults by states and banks as reason to press ahead with stabilizing measures, decision-makers could have opened processes of debt renegotiation, using the moment to challenge tenets about the inviolability of debt.[80] Acknowledging in this way some of the longer-term problems involved, they could have asked *what kind* of socio-economic order to pursue rather than how to maintain the status quo, and worked within existing political norms rather than sought to circumvent them in the name of rapid response. A default within the Eurozone need not have put its existence at stake[81]—and its existence in any case did not need to go unquestioned. To approach this as the politics of emergency is to approach it as a chosen mode of rule, not just an unavoidable crisis response. Such a perspective allows us to see the idea of the sudden onset of malaise, and the implied moral warrant for fast, a-normative responses, as contestable claims rather than as constitutive features of the situation. It shifts our focus from 'integration-through-crisis' to the methods used and the kinds of integration produced.

One of the central insights of studies of emergency rule is the importance of studying the irregular moments—of examining apparently temporary institutional arrangements as research objects in their own right rather than as stepping-stones in a process of completion. Improvised arrangements in the EU are too often approached as formal-arrangements-in-waiting, to be absorbed into a version of the existing framework. By contrast, it is often crucial that, for a period at least, they lie outside that framework, and typically for an uncertain timespan. It is this that can ease the passage of controversial measures. That they are rationalized by appeal to exceptional circumstances may be one of their most significant features, not just a detail of their timing.

One needs to examine in other words the *logic of response*, not just the circumstances to which actors are responding. This is important for a critical assessment of these practices and the legitimacy they claim for themselves, as well as to understand how they support or interfere with such ideals commonly vested in the EU as the rule of law, technocracy and democracy. It is equally important for understanding the responses these practices provoke. To view the EU through the lens of emergency rule is to contribute to a critical scholarship of the EU that has emerged in legal studies, while alerting us to things that go under the radar if

[80] Sandbu (*Europe's Orphan*), pp. 8–9; cf. Mody (*EuroTragedy*), p. 15.
[81] Sandbu (*Europe's Orphan*), p. 71.

we look only at the letter of the law.[82] It is to look at something more focused, but also more troubling, than a general notion of crisis-led experimentalism.

A second reason to adopt this view is more empirical. To think in terms of emergency rule is to highlight dynamics that cut across the major existing theoretical perspectives. The principal axis of EU scholarship remains the opposition between accounts emphasizing supranational and intergovernmental institutions.[83] Both are compelled to underplay important dynamics to maintain their place in the theoretical conversation. Supranationalist accounts will tend to be quiet on the use of non-EU frameworks in decision-making, as well as the place of national and sub-national agents in realizing the measures adopted. Intergovernmentalist accounts will tend to overlook the role of the ECB and its capacity to expand its own powers, the development of the Commission as a discretionary agent, as well as the increasing inequality of member-states. Arguably both miss therefore the rise of executive activism at *all* political levels, and the dynamics to which this responds.

To the extent that the field as a whole features both perspectives, the deficiencies of each do not go unnoticed, and indeed may be a source of productive tension. But there are also blind-spots they share. Since its initiation, the EU has tended to be studied as a set of institutions, each with its own agenda and interests. This institutional ontology appears in the analysis of the Euro crisis, with a prime analytical goal being to identify which institutions are the 'losers' and the 'unexpected winners'.[84] Taking seriously such formations as the Troika suggests though the importance of examining ties that cut across institutions and divide them from within. The emergency framing alerts us to the shape-shifting constellations and tendencies towards de-institutionalization that can be expected once exceptional circumstances are called. It alerts us to how, under such conditions, a cross-cutting set of agents may respond to cross-cutting concerns. When the stakes are considered high, the institutional ontology can be misleading.

Closer to the perspective I am advocating are those third positions in EU scholarship, beyond supranationalism and intergovernmentalism, that focus on

[82] E.g. Christian Joerges and Christian Kreuder-Sonnen, 'European Studies and the European Crisis: Legal and Political Science between Critique and Complacency', *European Law Journal* 23 (1–2) (2017), pp. 118–39; Kilpatrick ('On the Rule of Law'); Chalmers et al. (*End of the Eurocrats' Dream*); Augustin Menendez, 'The Existential Crisis of the European Union', *German Law Journal* 14 (5) (2013), pp. 453–526; Nicole Scicluna, *European Union Constitutionalism in Crisis* (London: Routledge, 2015); Suntrup ('From Emergency Politics . . . ').

[83] E.g. Chris Bickerton, Dermot Hodson, and Uwe Puetter, 'The New Intergovernmentalism: European Integration in the Post-Maastricht Era', *Journal of Common Market Studies* 53 (4) (2015), 703–22; Frank Schimmelfennig, 'European Integration (Theory) in times of crisis: A comparison of the euro and Schengen crises', *Journal of European Public Policy*, 25 (7) (2018), pp.969–89; for a useful critical perspective: Vivien Schmidt, 'Rethinking EU Governance: From "Old" to 'New' Approaches to Who Steers Integration', *Journal of Common Market Studies* 56 (7) (2018), esp. p. 1546.

[84] Michael Bauer and Stefan Becker, 'The Unexpected Winner of the Crisis: the European Commission's Strengthened Role in Economic Governance', *Journal of European Integration* 36 (3) (2014), pp. 213–29.

tendencies that play out across multiple institutions—the rise of executive power and discretion, of a European power elite, of 'tangled governance', or the appeal to 'independence' as an organizing and legitimizing principle.[85] Each emphasizes tendencies that cannot be localized on particular sites of authority. By paying closer attention to the emergency logic with which such practices are so often entwined, one gains insight into the factors that drive them and the rationalities on which they depend. The appeal to exceptional circumstances is, I suggest, a constitutive element in these long-term tendencies, not just a context in which they appear.

Conclusion

In the EU crisis politics of the 2010s, the figure of the Troika loomed large. Though on one level it was no more than an icon, a receptacle for public emotion that verged on the cartoonish, on another it conveyed something deeper. It encapsulated a certain kind of politics, based on ostensibly impermanent, unconventional arrangements for defined purposes, governed by an agenda of speed and urgency, and propelled by trans-institutional executive power. These themes were central to the decision-making of the period, and taken together amount to a specific mode of governing, usefully grasped with theories of emergency rule. Varying in important ways from the classical models, the EU case also gives a starting-point for thinking about the specificities of emergency rule in the transnational context.

If emergency-based exceptionalism were just about problem-solving, one might celebrate such moves as the exercise of leadership. One might acclaim them as executive power stepping forward to take responsibility under the pressure of events. But one reason for widespread suspicion of the Troika and its partners was that such responses exceed functional adaptation, favouring some solutions over others. Indeed, 'events' themselves are what powerful decision-makers make of them, and it is important to consider further the commitments bound up in their identification. Rather than asking how certain challenges are overcome, it is worth asking what one can learn about the outlook of executive power from the kinds of challenging situation it prioritizes.

The following two chapters develop this thought by turning to the historical background. They seek to contextualize the patterns described, investigating what drives emergency politics in the EU and examining for how long it has been apparent.

[85] Curtin (*Executive Power*); Henning (*Tangled Governance*); Vauchez (*Democratizing Europe*). On European power elites: Niilo Kauppi, and Mikael R. Madsen (2013) (eds), *Transnational Power Elites: The New Professionals of Governance, Law and Security* (London: Routledge). Following C. Wright Mills, one of the distinguishing features of a power elite is that it cuts across institutional divisions: it is shaped by informal ties and shared outlooks for which formal affiliations are a weak guide.

2
Structural Vulnerabilities of the Transnational Sphere

In his classic biography of the 'first statesman of interdependence', Francois Duchêne describes the working methods of Jean Monnet as breathless and purposefully so. 'The atmosphere of permanent emergency he generated (and to which quite a few strongly objected) drove people of different and stubborn traditions to overwork together and, almost without knowing it, to develop a corporate identity and pride.'[1] This was an energetic, irregular, and highly personalized mode of operation. Formal rules and divisions of role were things Monnet wished to avoid for Europe's new executive, the High Authority of the Coal and Steel Community. The slow procedures of bureaucracy exemplified by the contemporary nation-state were to be avoided—not least because the task of the moment was to loosen Europe's dependence on the nation-state model. It was in the inverse of statist proceduralism that the transnational should define itself.

Monnet's vision of an intentionally disorganized, open-ended style of rule did not survive as the dominant form in the emerging European Community. As Duchêne would observe, 'at first, the atmosphere of permanent crisis generated a certain creative tension, but it was not possible to run a major institution like that for long'.[2] As incoming head of the European Commission, Walter Hallstein pursued a more formal arrangement and sought to ground the institution's activities in legal structure. A more bureaucratic image of the EU would come to dominate—not so much the rule of persons, informal and discretionary, as something closer to the rule of rules.

But Duchêne's insight into the working methods of the early Community invites us to examine whether the seeds of emergency rule were not present from the beginning. Was there a natural affinity between the structure of the transnational realm and the 'atmosphere of permanent emergency' that characterized the Monnet era? Were there long-term factors conducive to a politics unbound by politico-legal norms, shaped by ideas of exceptional circumstance and driven by executive discretion? And if such tendencies could be overlooked

[1] Francois Duchêne, *Jean Monnet: The First Statesman of Interdependence* (New York: W. W. Norton & Company, 1994), p. 239.
[2] Duchêne (*Monnet*), p. 240.

Politics of Last Resort: Governing by Emergency in the European Union. Jonathan White, Oxford University Press (2020). © Jonathan White.
DOI: 10.1093/oso/9780198791720.001.0001

for significant periods of European integration, what has led to their renewed visibility in the EU's more recent history?

In this chapter I examine some of the ways in which the transnational sphere makes the politics of emergency a persistent temptation. I look first at how the EU has emerged in close connection with a particular set of objectives that have tracked its development. Since its early days, European integration has involved the creation of institutions whose rationale is to advance and maintain a certain body of policies, understood as ways to advance broader goals of peace and prosperity. Those associated with the 'four freedoms' of the common market, to do with the movement of goods, capital, services, and labour, have long been the core, gradually supplemented by others. While the policy commitments have expanded over time, relatively constant has been the way political structures have been shaped around particular ends—'form follows function', as the phrase has it.

As I argue, the effect is that policy commitments have always had basic priority over procedural arrangements in EU politics. Rather than self-standing entities that can be put to different ends, broadly on the model of the modern state, one sees institutions and procedures evolving with the policies, and liable to be sidestepped should they fail to serve those ends. The centrality of certain defining goals to the integration process means cross-cutting considerations are liable to be marginalized when a trade-off is held to be necessary. There is an instrumentalist orientation to the transnational polity that makes its arrangements precarious when the going gets tough. Maintaining certain policies becomes synonymous with maintaining the association, while challenges to them put the association in doubt. A basic distinction between order and disorder is thereby embedded, one that institutional representatives can invoke in numerous ways to rationalize discretionary moves.

A non-hierarchical constitutional structure does little to inhibit these manoeuvres, indeed arguably gives further reason for them. The diffuse institutional structure of the transnational realm, comprising multiple contending sites of power, is an additional source of vulnerability to the politics of emergency. To the extent that the EU has been built around certain purposes, it has been dependent on the coordination of multiple agents for the achievement of its defining goals. A long-standing idea in constitutional thought is that the diffusion of power across many institutions is prone to create paralysis, especially when the distribution of roles is weakly underpinned constitutionally. Arriving at common policies and enforcing them becomes the work of many hands. Accordingly, pressure may build for a re-centralization of power when considerations of urgent necessity can be invoked. The ideas and practices of emergency are a way to galvanize action, coordination, and innovation across a diverse and potentially recalcitrant institutional field—as perhaps Monnet himself was already aware.

These are structural vulnerabilities to emergency rule, both in the sense that they are long-standing and that they follow from distinctive features of the transnational realm. They were present from the earliest phases of European integration, and occasionally gave rise to periods of rule that foreshadow more recent events—to periods of emergency politics 'in a minor key', as one might call them. The creation of the European Council in the 1970s offers a useful example that is the focus of the final section. Tendencies to emergency-led executive discretion could largely be contained, however, when geopolitical and socio-economic conditions were accommodating, and when bureaucratic and legal forms of authority enjoyed a certain prestige. For much of the later twentieth century, these conditions endured, and an image of the EU as a rule-bound association could hold sway. But as Chapter 3 will show, historical changes both in the wider societal environment and in the ethos of executive power itself would subsequently reinforce these underlying vulnerabilities, making their presence increasingly telling.

Transnational Politics and the Instrumentalist Attitude

European integration emerged out of a project of advancing cross-border economic coordination. This, very roughly, is a point of consensus in EU scholarship. One school of interpretation sees this as a process driven by the defined objectives of member-state governments as shaped by domestic groups. For liberal intergovernmentalists, the founding policies of European integration were essentially the sum of what state representatives could agree was in their shared economic interest.[3] Another school emphasizes that core goals of coordination in certain sectors tended to spill over into related fields, creating an expansive logic of integration. Neofunctionalists put an accent on the open-endedness of the project of economic coordination.[4] A third, Marxist-informed body of work notes that the arising coordination aligned with the interests of major capitalist producers. Class interests were well served.[5] What is common to these accounts, despite their differences, is the centrality of economic purposes in the initiation of what would become the EU.

[3] Alan Milward, *The European Rescue of the Nation-State* (London: Routledge, 1992); Andrew Moravcsik, *The Choice for Europe: Social Purpose and State Power from Messina to Maastricht* (Ithaca: Cornell University Press, 1998).

[4] Ernst B. Haas, *The Uniting of Europe: Politics, Social and Economic Forces, 1950–1957* (Stanford: Stanford University Press, 1968); Philippe C. Schmitter, 'A Revised Theory of Regional Integration', in L.N. Lindberg and S. A. Scheingold (eds), *Regional Integration: Theory and Research* (Cambridge, MA: Harvard University Press, 1971), pp. 232–64; Philippe C. Schmitter, 'Neo-neo-functionalism?', in Antje Wiener and Thomas Diez (eds), *European Integration Theory* (Oxford: Oxford University Press, 2004), pp. 45–74.

[5] E.g. Ernest Mandel, 'EEC: The Common Market in Crisis' *Inprecor* 2 (June 20, 1974), pp. 14–18.

The institutions of the emerging order were designed with the fulfilment of economic tasks in mind. They were conceived, in the first instance, as ways to achieve these. One sees this already in the Paris Treaty (1951), where the High Authority and Assembly were presented as mechanisms overseeing the functional ends of the Coal and Steel Community, and in lasting form once economic coordination decisively became associated with the project of constructing a common market with the Treaty of Rome (1957) founding the European Economic Community. Recasting the institutions as the Commission and Parliament, the Treaty made clear that they existed to advance a certain set of goals, centred on removing obstacles to the movement of goods, persons, services, and capital.[6] The policy goals were defined first, and the institutions as means to advance them. It was the making of a sphere in which institutions were closely associated with particular substantive ends. Much less successful were the parallel, federal initiatives intended to establish a more wide-ranging institutional framework that could be put to broader ends.[7]

The outcome was a political order that was initially designed for, has subsequently developed in accordance with, and also finds its rationalization in, the fulfilment of policy goals to do with the market-based economic coordination. The EU can be viewed in this sense as a purposive association, or what Hans-Peter Ipsen once called a *Zweckverband*.[8] An instrumentalist rationale for European integration has long been a pronounced current in the EU's political and legal discourse.[9] Turkuler Isiksel has recently introduced the idea of 'functional constitutionalism' to describe how functional ends to do with market-making have shaped the EU's structure and the legitimacy claims made of it, both in its early years and its later development. As she writes, 'the EU's legislative mechanisms, its extensive regime of individual rights, citizenship provisions, and the doctrinal

[6] Art. 2 outlines the general economic rationale of the EEC; Art. 3 outlines the core policies to be pursued 'for the purposes set out in the preceding Article'; Art. 4 then states 'The tasks entrusted to the Community shall be carried out by the following Institutions...'. Further substantive goals are elaborated throughout the Treaty: see e.g. Art. 104 on the requirement for member-states to 'pursue the economic policy needed to ensure the equilibrium of its overall balance of payments and to maintain confidence in its currency, while taking care to ensure a high level of employment and a stable level of prices'.

[7] For the classic statement: Altiero Spinelli and Ernesto Rossi, 'The Manifesto of Ventotene for a Free and United Europe' (1941). (https://www.cvce.eu/content/publication/1997/10/13/316aa96c-e7ff-4b9e-b43a-958e96afbecc/publishable_en.pdf), advocating 'the creation of a solid international state', with intrinsic value as signalling the end of the artificial division of Europe into nation-states.

[8] The term 'Zweckverband funktioneller Integration' comes from Hans Peter Ipsen, *Verfassungsperspektiven der Europäischen Gemeinschaften* (Berlin: de Gruyter, 1970). For discussion: Joerges in Chalmers et al. (*End of the Eurocrats' Dream*), p. 312; also Damian Chalmers, Gareth Davies, and Giorgio Monti (2014), *European Union Law: Text and Materials* (Cambridge: Cambridge University Press), p .60.

[9] See e.g. land-mark ECJ judgements such as *van Gend en Loos* (5th February 1963), which treats the Community as defined by its purpose of establishing a Common Market: cf. Damian Chalmers and Luis Barroso, 'What *Van Gend en Loos* Stands For', *International Journal of Constitutional Law* 12 (1) (2014), pp. 105–34.

structure of its legal system are conditioned by and tailored to this purpose. Its constitutional features draw their authority from the functional imperatives of creating and maintaining an economic union, understood ... in the broad sense of a policy space that encompasses a single market, monetary union, exhaustive fiscal coordination, and regulatory powers ranging from consumer protection to environmental standards.'[10] The EU and its institutions emerged out of the pursuit of certain policy goals.

Of course, this orientation to the purpose of market-making was only part of the story for the principal agents involved. European integration was rarely considered reducible to functional considerations alone: ideological influences in the early decades tended to locate market-making within a broader scheme. This larger perspective tended to include more general ideas of peace, prosperity and the good life, often combined in the manner of eighteenth-century ideas of 'doux commerce', such that good commercial relations were expected to bring harmonious and enabling relations more generally.[11]

At least two bodies of thought tend to be highlighted as sources of inspiration in the EU's formative years. There were, first, various strands of liberal-economic thought with some purchase on senior officials, in which building the common market was connected to a project of restructuring state-market relations. Ordoliberalism has gained particular recent attention in this respect, though in this period it shaded into a number of other strands too.[12] For these liberal-economic advocates of European integration, transnational coordination was a way to pursue the depoliticization of economic relations and to protect entrepreneurs from unwelcome forms of state intervention and monopoly capitalism. It was, in the language of Ordoliberalism, a matter of cementing certain market-enabling policies known collectively as the 'economic constitution', so that liberty, free enterprise and virtues such as responsibility could flourish.

[10] Turkuler Isiksel, *Europe's Functional Constitution: A Theory of Constitutionalism Beyond the State* (Oxford, Oxford University Press, 2016), p. 7; cf. Curtin (*Executive Power*), p. 30.

[11] On 'doux commerce': Albert Hirschman, *The Passions and the Interests: Political Arguments for Capitalism Before Its Triumph* (Princeton: Princeton University Press, 1996); as it bears on European integration: Jonathan White, 'Europe and the Common', *Political Studies* 58 (1) (2010a), pp. 104–22, p. 107.

[12] For historical and analytical overviews of Ordoliberalism and the EU: David J. Gerber, 'Constitutionalizing the Economy: German Neo-liberalism, Competition Law and the "New" Europe', *The American Journal of Comparative Law* 42 (1994), pp. 25–84; Viktor J. Vanberg, 'The Freiburg School: Walter Eucken and Ordoliberalism', Paper 04/11 (Freiburg: Walter Eucken Institut, 2004); Werner Bonefeld, *The Strong State and the Free Economy* (Lanham: Rowman & Littlefield, 2017); Josef Hien and Christian Joerges (eds), *Ordo-Liberalism, Law and the Rule of Economics* (Oxford: Hart, 2017); Thomas Biebricher, 'Europe and the Political Philosophy of Neoliberalism', *Contemporary Political Theory* 12 (2013), pp. 338–75; Quinn Slobodian, *Globalists: The End of Empire and the Birth of Neoliberalism* (Cambridge MA: Harvard University Press, 2018), ch. 6. Challenging 'overstated' claims for Ordoliberalism: Andy Storey, 'Authoritarian Neoliberalism in Europe: The Red Herring of Ordoliberalism', *Critical Sociology* (online early, 2019).

The transnational sphere was attractive for a number of mid-century Ordoliberal thinkers since it offered a new front on which to push for an economic constitution, one relatively detached from the influence of mass publics. Here the possibilities were particularly open for expert networks to take shape and to use problem-solving capacity as a warrant for new powers. More promisingly than the nation-state alone, it offered the potential for a purposive association that could advance and embed a certain model of economic order. And for Ordoliberals, the market-building project was about far more than the functional pursuit of efficiency. It was about cementing the basis of a moral order founded on a certain vision of the good life.[13]

A second ideological influence on early European integration was Christian Democratic thought. Its partisans had a notable presence in the governments of the community's founding members, and arguably were especially successful in shaping European integration in their image.[14] Again, this was an orientation in which the construction of a common market was linked to a wider agenda—not only to an economic vision but to an idea of Europe as a religio-civilizational entity. Pursuing economic coordination amongst six continental states was intended to strengthen Christian Europe and its socio-economic order in the context of Cold War hostilities. It was intended to consolidate the 'social market economy', based on private property and a limited redistribution of wealth, that was understood as the foundation of a Christian social order. As recent scholarship on Christian Democracy has emphasized, this was a perspective in which public authorities tended to be rationalized in purposive terms, as '"instruments" of the body politic, created for the purpose of fulfilling specific tasks required in the overarching pursuit of the temporal common good'.[15] A transnational order built in the Christian Democratic image was one in which institutions would be regarded as functional to the achievement of certain ends.

A certain set of policies to do with market-making was thus where a diversity of political agendas overlapped. These policies achieved pre-eminence not just because they promised material dividends for those in power, but because they were the point of convergence for wider ideological projects. Adherents of liberal-economic and Christian Democratic thought generally supported the European market-building process for different reasons. For each, it was an extension of their own normative projects. But what they shared was a purposive outlook on

[13] Bonefeld (*Strong State*), p. 83. A recent study on transnational variants of Ordoliberalism—or 'Ordoglobalism'—notes there were disagreements concerning the attractiveness of an economic constitution on a European scale. Some, like Wilhelm Röpke, saw it as compromising a properly universalist orientation, while others, notably Hans von der Groeben and Ernst-Joachim Mestmäcker, treated it as an advance on statist approaches: see Slobodian, (*Globalists*), p. 183, pp. 207ff.

[14] Wolfram Kaiser, *Christian Democracy and the Origins of European Union* (Cambridge: Cambridge University Press, 2007); Carlo Invernizzi Accetti, *What is Christian Democracy?* (Cambridge: Cambridge University Press, 2019).

[15] Invernizzi Accetti (*Christian Democracy*), pp. 256–7.

European integration in which its value was closely tied to its capacity to advance a certain socio-economic order. Transnational institutions were formed not so much out of a principled commitment to internationalism. They were formed to advance and embed certain policies, and should they fail to contribute to this end, or hinder it, they would cease to enjoy the same value.

This instrumental character has arguably continued to feature over the course of European integration, as new policy goals have been added to the original configuration, and with them new institutions. Even as the EU has expanded in policy and territorial range, the same basic idea of the EU as an instrument to perform delimited tasks has been retained. Monetary union would come to the fore as a defining objective of EU politics as the common-market programme neared completion in the 1990s. The emergence of the ECB, as an agent defined by a certain purpose (price stability), contributed to the reproduction of this conception, as did the creation of many other task-specific agencies in the EU galaxy. Form has continued to follow function: indeed, in view of some tendencies we shall examine in the next chapter, if anything this instrumental logic has become more pronounced.

As I now want to suggest, the centrality to the EU of a certain set of policy goals creates a structural vulnerability to the politics of emergency. In a number of interconnected ways, it makes extraordinary discretion under the sign of emergency a persistent possibility, even if its exercise is deferred.

The Primacy of Ends

When an association is constituted to pursue certain policies, and its legitimacy is grounded in their successful pursuit, a basic asymmetry is written into its activities. The ends around which the union is built acquire a distinct prestige. Honouring the common-market programme outlined in the Treaty of Rome has long been the guiding concern for the EU's supranational institutions. Actions taken in the name of a purposive association tend to be justified as acts of fidelity to the purposes said to define it.[16] By contrast, concerns not defined as part of its rationale can easily become second-order concerns, subordinated to the pursuit of the former. At least under certain conditions, when it is felt that the goals of the association are under threat and that hard choices must be made, protecting these goals at the expense of competing concerns will predictably seem the appropriate course. When general agreement on a certain set of ends is supposed, and on these alone, single-mindedness in their pursuit can be expected.

[16] Jonathan White, 'Revisionism as a Logic of Institutional Change', *European Law Journal* 23 (5) (2017), pp. 406–16.

At the level of administration, constituting governing arrangements around a delimited set of policy goals entails a separation of spheres of authority. Officials in supranational institutions are made responsible for specific objectives rather than public policy in the round. In those spheres they are expected to be objective and 'independent'.[17] But overall they are expected to take a sectoral view, and are given powers to address certain matters and not others.[18] This sectoral quality may be part of the appeal of such an order for those who believe an equilibrium can be maintained, such that activities in one sphere complement rather than intrude on those in another.[19] But even where the division is sustainable in this way, the price is likely to be an administrative order whose component parts display a certain indifference to the concerns of other sectors.

By prioritizing the market-making process, with its emphasis on the free movement of goods, capital, services, and labour, and by subsequently appending additional goals of economic coordination such as monetary union, the pioneers of European integration fostered a political space in which competing considerations would present themselves as obstacles to core business. Policy goals external to those forming the core purposes of the association, e.g. economic considerations difficult to subsume under the heading of market-making, would be some of the concerns susceptible to marginalization.[20] Likewise, and importantly for our discussion, the range of concerns liable to be diminished would include political and legal norms, e.g. to do with democratic control, which are non-essential to the association's instrumental rationale. To quote Isiksel again, the EU order 'posits economic objectives such as productivity, competitiveness, and higher standards of living as superordinate goods whose pursuit takes precedence over democratic self-rule by citizens, which is enshrined in the EU's institutional architecture only to the extent that it will minimally hinder the provision of these superordinate goods'.[21] A purposive association holds the enduring potential to sacrifice concerns incidental to its rationale.

Formally, the EU has long recognized procedural norms which constrain the pursuit of its core economic goals. Principles of democracy and the rule of law are acknowledged as basic features of the political life of its member-states.[22]

[17] Vauchez (*Democratising Europe*).

[18] See Invernizzi Accetti (*Christian Democracy*), p. 137: 'Since supranational instances of government are justified as logical extensions of the principle of subsidiarity beyond the sphere of the nation-state, they are understood as additional layers of *partial* authority, with a set of very specific functions, defined by the concrete problems nation-states are assumed to be incapable of adequately addressing at their level.'

[19] Ipsen (*Verfassungsperspektiven*).

[20] Fritz Scharpf, *Governing in Europe: Effective and Democratic?* (Oxford: Oxford University Press, 1999).

[21] Isiksel (*Functional Constitution*), p. 91; cf. p. 18.

[22] See Article 6 (1) TEU, which—following the Amsterdam Treaty—speaks of the EU as 'founded on the principles of liberty, democracy, respect for human rights and fundamental freedoms, and the rule of law'.

At the transnational level, the Community Method and conventions to do with the exercise of national vetoes amount to procedural norms within which the pursuit of policy goals is expected to unfold. There have been substantial periods in the history of European integration when decision-making has been broadly consistent with these norms. The widespread image of the EU as a community of law, indeed as sometimes overly committed to legal and bureaucratic methods, is testament to their potency as images of how the EU is supposed to work.

But the status of such norms in a political order oriented to specific policy ends is always likely to be secondary in the last instance. When institutional structures and the procedures associated with them are at odds with the association's defining goals, they will predictably be deemed expendable. The inclination will be to sacrifice or suspend them, and to create new arrangements better attuned to the pursuit of those ends held 'indispensable'.[23] Too firmly constitutionalized a structure is liable to inhibit the pursuit of the association's defining goals.[24] Procedures will be tailored to what the policies seem to demand—unlike the ideal-typical modern state, where procedures are expected to constrain the pursuit of policy ends.

The EU is often characterized as a highly judicialized order, shaped by a powerful and interventionary Court of Justice (CJEU). It would be misleading however to see the CJEU as the countervailing force that upholds the EU's constitutional norms. The role of the CJEU has tended to be more targeted. Its focus has been on ensuring national compliance with EU decisions, more than on overseeing and enforcing the procedures of EU decision-making itself.[25] The Court's role in emergency decision-making is something Chapter 7 returns to, but the point to note here is that the existence of an activist Court has not countered the EU's basic devotion to certain substantive policies and the vulnerability to emergency rule this entails.

How do such observations square with the common assessment that the history of European integration has been a project of *restricting* executive discretion? This

[23] On themes of indispensability in the founding discourses of European integration: Claudia Schrag Sternberg, *The Struggle for EU Legitimacy, 1950-2005* (Basingstoke: Palgrave Macmillan, 2013), pp. 16ff.

[24] A theoretical statement can be found in David Mitrany's work: 'the function, one might say, determines the political instrument suitable for its proper activity, and by the same means provides for its reform at every stage. Not only is there in all this no obvious need for constitutional division of authority and power, but any except the most general formal rules would embarrass the working of the system.' (David Mitrany, 'A War-Time Submission', in *The Functional Theory of Politics* (London: Martin Robertson, 1941/1975), p. 118). On Mitrany's thought and its wider influence: Or Rosenboim, *The Emergence of Globalism: Visions of World Order in Britain and the United States, 1939–1950* (Princeton: Princeton University Press, 2017).

[25] The focus of EU legal scholarship also tends to be on national compliance rather than the conduct of EU institutions. On 'systemic' departures: Armin von Bogdandy and Michael Ioannidis, 'Systemic Deficiency in the Rule of Law: What It Is, What Has Been Done, What Can Be Done', *Common Market Law Review* 51 (2014), pp. 59–96.

is a recurrent theme of both political-science and legal scholarship on the EU.[26] It is also one of the grounds on which a congruence tends to be observed between European integration and liberal-economic thought of the Ordoliberal kind. Analyses of Ordoliberalism tend to present it as an ideology strongly attached to legal methods and rules. Many of its more influential proponents wanted to limit the scope for executive discretion and to forge a rule-bound order.[27] Is such an outlook compatible with the prioritization of substantive ends over procedural norms?

The two are quite compatible when the purpose of rules is to underpin a certain policy orientation. With their emphasis on an economic rather than political constitution, Ordoliberals were more interested in constraining the kinds of economic policy that political agents could pursue than the ways in which they might initiate and enforce it. A strong commitment to upholding rules of policy may invite great latitude in the procedures by which this is achieved. Early Ordoliberals famously emphasized the importance of a strong state that was, as it were, legally unencumbered to do what economically it must.[28] Here was a tradition that reserved a place for unbounded executive action in the service of its preferred market-based order.[29]

The observation carries added relevance because by embedding certain economic policies in the EU order, its makers were arguably planting the seeds of later instability, and therefore the conditions in which extraordinary action might seem appropriate. Note in particular how the centrality of certain policies to the integration project raises the ongoing challenge of responding to changing conditions. To build an association around certain agreed ends is to bake these into its governing structure. Dieter Grimm has emphasized how substantive policy commitments acquire a quasi-constitutional status in EU politics. The *acquis* of EU law is a body of policy commitments that, being written into treaty law, are hard to revise or reverse, amounting to a pattern of what Grimm calls 'over-constitutionalization'.[30] One of the effects of embedding policies in this way is to make decision-making prone to rigidity. To the extent that socio-economic

[26] E.g. Christopher J. Bickerton, *European Integration: From Nation-States to Member-States* (Oxford: Oxford University Press, 2012); Tuori & Tuori (*Eurozone Crisis*).

[27] Franz Böhm endorsed an 'automatically functioning coordination system' that would 'severely limit political discretion': Böhm, 'Privatrechtsgesellschaft und Marktwirtschaft', *ORDO Jahrbuch für die Ordnung von Wirtschaft und Gesellschaft* 17 (1966), pp. 75–152.

[28] Hence the familiar depiction of it as tending towards authoritarian liberalism. The historical link between early strands of Ordoliberalism and Carl Schmitt's work is discussed in Bonefeld (*Strong State*).

[29] On the complex relation between rules and discretion in Ordoliberalism: White ('Revisionism'); David M. Woodruff, 'Governing by Panic: The Politics of the Eurozone Crisis', *Politics & Society* 44 (1) (2016), pp. 81–116.

[30] Dieter Grimm, *The Constitution of European Democracy,* (Oxford: Oxford University Press, 2017).

circumstances and societal preferences evolve over time, the shared policies that bind the association in one era are likely to seem less apposite in another.

Legal theorists in the American pragmatist tradition used to speak of a dialectic of rules and discretion. As they argued, when policy-making becomes increasingly regularized, so the pressure for occasional discretionary intervention increases, which in turn creates new pressure for regularization. As societies evolve, in their socio-economic practices and political ideals, a mismatch may emerge between the context in which policy commitments crystallized and the context in which they are later pursued, with resulting pressure for their revision.[31] In a political association such as the EU, built around a core of rather detailed policy commitments, such dynamics are all the more pronounced. Adaptation becomes the persistent challenge.

Sometimes it may be possible to meet this challenge by gradually refining the shared purposes that define the association, consistent with neofunctionalist expectations about European integration being a process of incremental change.[32] But what happens when more far-reaching reorientations of policy are sought? What happens for instance when the advance of financial capitalism changes what it means to pursue economic coordination founded on a common market and (later) common monetary policy? Redefining the goals of a purposive association can be contentious, since such moves may put in doubt its very rationale. There are also likely to be limits to how far such a process of redefinition can plausibly be taken. When these more challenging contexts arise, the temptation for governing agents will be to invoke ideas of crisis to authorize radical shifts. It will be to cast the situation as an exceptional one warranting an extraordinary response, invoking necessity to rationalize changes that would otherwise be difficult.

Because, in other words, the purposive association struggles to abandon goals that form its basic rationale, as well as to adopt new goals unrelated to its founding aims, and because these commitments must contend with a dynamic environment, extraordinary manoeuvres are a persistent possibility. Conceiving transnational order in purposive terms encourages, we might say, any challenge not easily absorbed within the existing framework to be cast as the threat of disorder, to be staved off by vigorous means. Entrenching a certain set of policy ends makes thinkable the possibility of far-reaching discretion unconstrained by the usual norms of politics and political justification.

In the scholarship on the EU's crisis politics of the 2010s, the ECB's overriding goal of ensuring price stability is often highlighted as key to understanding its extraordinary actions. This defined policy goal embedded an aversion to debt

[31] See Roscoe Pound, 'Justice According to Law', *Columbia Law Review* 13 (8) (1913), pp. 696–713; Morris R. Cohen, 'Rule vs Discretion', *Journal of Philosophy, Psychology and Scientific Methods* 11 (1914), pp. 208–15.

[32] I have referred to this elsewhere as *revisionism* (White, 'Revisionism').

restructuring, seen as liable to weaken the currency and encourage speculation, and prompted the search for alternatives. It encouraged the ECB head to press for the adoption of austerity measures in indebted states, using anti-democratic methods to do so (notably secret, threatening letters to heads of state), and led ultimately to the commitment to buying the bonds of indebted states. Because such moves were themselves dubiously compatible with the ECB's mandate, agents emphasized the exceptional circumstances facing the Eurozone as a whole as a way to legitimize these shifts.[33] My suggestion is that what these particular actions express is a more general feature of how the EU has developed. They reflect how institutions have come to serve particular goals, and how the corresponding primacy of ends can prompt recourse to emergency politics.

If the disposition to see transnational institutions in these instrumentalist terms creates an affinity for emergency rule, it is arguably no more than that. Much of the history of European integration has proceeded without recourse to emergency logic, for reasons we shall say more of. But while European integration could under certain conditions appear the very opposite of the discretionary methods of emergency politics, its structural connection to the policy regimes associated with the common-market programme has entailed the latent prospect of extraordinary responses in difficult situations to uphold the socio-economic order to which its adherents are committed.

Decision without Hierarchy

A defining feature of the pursuit of common policies at the European level has been the avoidance of formal relations of hierarchy and the creation of a sovereign authority. Nothing like the Hobbesian image of a centrally governed Leviathan has been adopted as the means to advance policy ends in the transnational context. The EU is not a federal state. Coordination has tended to rely on customary patterns of consultation across institutional settings rather than on top-down processes underpinned by credible sanctions. This follows in part from the primacy of certain policy goals to the overall enterprise. When public authority is geared to the efficient pursuit of certain ends, elasticity in the relations between institutions can be advantageous.

It follows also from the ideological influences noted. This non-hierarchical structure is arguably another legacy of Christian Democratic involvement. Scepticism towards the principle of sovereignty, and a willingness therefore to challenge the centralized authority structures of the state while avoiding their re-emergence on a European scale, was a central current in Christian Democratic thought in

[33] Woodruff ('Governing by Panic'); Sandbu (*Europe's Orphan*).

the EU's formative years. The idea of a political centre of undisputed supremacy was viewed with suspicion on several grounds: as a challenge to a pre-political, Christian idea of natural justice, to the divine authority that underpinned this, and to the maintenance of transnational communities based on religion.[34] Political association on a European scale should be founded instead on the principle of *subsidiarity*, drawn from the Catholic Church, which entailed the dispersal of authority across multiple layers, supranational, national and subnational. Moreover, in the Christian Democratic view, the appropriate distribution of powers should emerge organically rather than be set by a written constitution. A formal codification of competences was one more mechanism by which positive law might interfere with the workings of natural law, and thus something to be deliberately avoided.

The influence of liberal-economic thought reinforced this preference for the diffusion of political power. In some strands, an anti-sovereign agenda was pronounced: the Hayekian argument for a European market programme was essentially that it would weaken the capacity of the state to make unwelcome interventions in the market. Authority at the European level would be limited in scope, and would be unable to use nationhood as a social resource for redistributive policies.[35] Ordoliberals too, though not averse to the strong state, saw European integration as a way to separate its powers across different levels, ensuring thereby that political authority was strong in its commitment to the economic constitution but weak in its capacity to favour the interests of certain producers.[36] The promise of law as a means to encase economic activity lay in its potential to straddle existing states without leading to the creation of a new centre.

What emerged was an order that was multi-centred, organized neither around the supremacy of existing states nor the emerging institutions of the supranational level.[37] Schmitt would probably have termed it a 'federation'.[38] Processes of coordination between multiple sites of authority, based on conventions of consultation rather than strict codification, have come to define the transnational sphere. And especially notable is the sheer number of agents across which these

[34] Invernizzi Accetti (*Christian Democracy*), Chapter 4.

[35] Friedrich von Hayek, 'The Economic Conditions of Interstate Federalism', in Hayek, *Individualism and Economic Order* (London: Routledge & Kegan Paul, 1952). Cf. Adam Harmes, 'Neoliberalism and Multilevel Governance', *Review of International Political Economy* 13 (5) (2006), pp. 725–49.

[36] Gerber ('Constitutionalizing...'), p. 37; Slobodian (*Globalists*), pp. 11ff.

[37] Neil MacCormick, *Questioning Sovereignty: Law, State, and Nation in the European Commonwealth* (Oxford, Oxford University Press, 1999).

[38] Carl Schmitt, *Constitutional Theory*, part IV: 'Constitutional Theory of the Federation' (Durham NC: Duke University Press, 1928/2008), p. 389: 'It is part of the essence of the federation...that the question of sovereignty between federation and member states always remains open as long as the federation as such exists alongside the member states as such'. P. 388: 'Such an intermediary condition necessarily leads to many conflicts, which must be decided.' In his view, the arrangement was workable if the states were sufficiently 'homogeneous' (p. 404); otherwise the federation's characteristic ambiguity concerning power would have to be resolved in an emergency.

processes of coordination unfold. Power has come to be diffused across a large number of executive institutions, including state authorities, the EU's core institutions, and a large number of functional agencies.[39] Decisions agreed at a European level also rely on national authorities for their implementation. Rather than controlled by a single agent, policies pursued through the EU depend for their execution on the coordination of many.

What one sees then is the weak regularization of how power is exercised combined with its diffusion across multiple sites. Again, I want to claim that these basic constitutional features open a space in which the politics of emergency can emerge.

To begin with, note how the diffusion of power across institutions undermines the capacity for decisive action. Progress is dependent on consensus across many sites. Sizeable obstacles to agreement may present themselves, including the differing agendas of supranational and intergovernmental representatives, of different national governments, and the ambivalent attitudes of national populations. Practices of mutual consultation can be successful if leading agents share a common ideological orientation (e.g. Christian Democracy) or at least a common understanding of the demands of economic coordination. But in the absence of such consensus, decision-making can be a convoluted process. Because the system is not designed to contend with disagreement and conflict, the prospect of paralysis arises.

One implication of this profusion of veto players is that it compounds the problem of how the association is to evolve with changing socio-economic circumstances. As the work of Fritz Scharpf has shown, decisions agreed at an EU level are challenging to revise in part because they are the work of many hands.[40] The broad outlines of policy are the precarious achievements of multilateral treaty negotiations, and major reorientations that would require treaty revisions will tend to be avoided. Here the constitutional structure of the EU reinforces the notion that it is a purposive association that adheres somewhat rigidly to a delimited set of goals. Certain types of policy initiative may be postponed indefinitely due to the difficulties of gaining agreement,[41] thereby storing up future problems. (We shall consider some of these in the next chapter when discussing the 'extreme conditions' to which emergency rule responds.) Where efforts to innovate *are* attempted, the agreement of many is needed. A treaty-based organization puts considerable power in the hands of those who negotiate treaties, i.e. executives, but for the most part they must seek parliamentary consent too. There will be the temptation—visible from the Euro crisis to the

[39] Curtin (*Executive Power*).

[40] Fritz Scharpf, 'The Joint Decision Trap Revisited', *Journal of Common Market Studies* 44 (4) (2006), esp. p. 848.

[41] Cf. Scharpf's account of the challenges facing 'positive integration' (Scharpf, *Governing in Europe*).

Brexit process—for governments to raise the stakes in a bid to gain acceptance, highlighting the prospect of chaos as a way to focus minds.

More immediately however, this diffusion of power creates an incentive to *bypass* the complexity of the system and concentrate power in the hands of the few when difficult situations arise. It invites a politics of executive discretion, improvised responses, and power reclaimed in the hands of individuals, especially given the formal constraints on doing so are weak. Impromptu decision-making forums allow the more powerful agents to apply pressure more effectively on the recalcitrant and to hasten their favoured course of action. Deadlines, a central coordinating device in a loosely structured association, can be used to add time pressure and raise the prospect of further disorder. The appeal to exceptional and pressing circumstances presents itself as a way to focus minds and unite disparate and divided agents behind a common decision. It is a way to legitimize a discretionary style of rule and to cajole other agents to coordinate and comply. Here again, Monnet seems to have understood something fundamental about the political utility of an 'atmosphere of emergency'.

The temptation to re-concentrate power in this way may be fostered further by those who see opportunities in such improvised formats to carve out a space for their own influence. Monnet's fondness for unconventional methods of decision-making was not just a managerial quirk applied to the running of one institution (the High Authority) but a way of overcoming resistance more generally and moving forward the integration project as a whole.[42] The diffuse structure of the transnational realm creates a special role for those who can bridge the diversity of institutions and link together the key protagonists.

Ultimately, there is also the option available to executives of going outside the EU framework to make arrangements based on international law. Faced with the protracted processes of consensual law-making, they may prefer ad hoc arrangements, legal in form but not part of a comprehensive legal framework. This abundance of options for rule-making has the effect of reinforcing executive discretion, enabling frameworks to be sidestepped when they become inconvenient. As long as national governments retain the capacity to make foreign policy, when thwarted by the EU's institutions they may seek 'coalitions of the willing' outside it. (This is consistent with the primacy of policy ends over procedures, since at any given moment it can be argued that the shared goals in question are best served by exiting existing arrangements.) The price of keeping executive

[42] Cf. Duchêne (*Monnet*), p. 347: 'For Monnet, a crisis was an opportunity. In a crisis, ministers, who are far too busy to act unless forced to do so, face choices which are as unpleasant if they fail to act as if they do. They are temporarily open to, even eager for, advice. "When ideas are lacking, they accept yours with gratitude—provided they can present them as their own". Monnet rarely had difficulty in envisaging a step which, in complex situations, might change the flow of action. He once said, without undue self-deprecation, "In crises most people do not know what to do. I do know." This attitude had the signal advantage, over generalized propaganda for "Europe", of being geared to life as politicians lead it.'

agents inside the system is likely to be agreements whose enforcement is selective, and which are easily supplemented by further ad hoc arrangements.

That there could be pressures for emergency rule endogenous to the structure of political institutions was something well understood by some classical theorists of emergency rule. Their observations formed with regard to the state and proto-state context anticipated points that can be made of the transnational setting. 'The worst defect weak republics can have is to be indecisive, so that all their decisions are taken out of necessity, and if any good comes to them, it comes through force of circumstance rather than through their own prudence.' These were the words of Machiavelli.[43] Republics based on the diffusion of power across multiple sites tended to work in 'slow motion', he argued, and if not supported by clearly defined procedures of authorization they would be vulnerable to improvisation in times of perceived need.[44] 'When a republic lacks such a procedure, it must necessarily come to ruin by obeying its laws or break them in order to avoid its own ruin.'[45] For Machiavelli, the answer was to make deliberate provisions for the centralization of power in individual hands at key moments. The republic of Venice was praiseworthy because it had 'reserved to a small number of citizens the authority to deliberate on matters of urgent need without consulting anyone else, if they are in complete agreement'.[46]

That institutional pluralism coupled with a weak constitution created scope for executive-led usurpations of power was a point recognized also by the American federalists. In *Federalist 20*, Madison described how in these conditions basic political norms could easily be circumvented in the name of crisis response. In a discussion of the Dutch confederacy intended to illustrate mistakes the US should avoid, he observed that 'in critical emergencies, the States-General are often compelled to overleap their constitutional bounds.... The treaty of Westphalia, in 1648, by which their independence was formerly and finally recognized, was concluded without the consent of Zealand. Even as recently as the last treaty of peace with Great Britain, the constitutional principle of unanimity was departed from. A weak constitution must necessarily terminate in dissolution, for want of proper powers, or the usurpation of powers requisite for the public safety. Whether the usurpation, when once begun, will stop at the salutary point, or go forward to the dangerous extreme, must depend on the contingencies of the moment. Tyranny has perhaps oftener grown out of the assumptions of power,

[43] Machiavelli (*Discourses*), I.38, p. 103.

[44] Machiavelli (*Discourses*), esp. I.34 (p. 95): 'the usual institutions in republics are slow to move, since no council nor any magistrate can undertake anything alone, for in many instances they need to consult one another; and, since time is wasted in coming to an agreement, the remedies for republics are very dangerous when they must find one for a problem that cannot wait'. See Geuna ('Machiavelli'), p. 227 for a gloss: 'Today one might say, with contemporary jargon, that republics are in fact polyarchies, but it is this institutional plurality—or polyarchy—that creates problems in times of emergency.'

[45] Machiavelli (*Discourses*), I.34, p. 95.

[46] *Discourses*, I.34, p. 195.

called for, on pressing exigencies, by a defective constitution, than out of the full exercise of the largest constitutional authorities.'[47] Again, for many of the American federalists the answer was to voluntarily centralize power in an orderly fashion, so that it did not have to be chaotically centralized in times of crisis. Hamilton's argument in *Federalist 70* for a unitary executive centred on a strong president was argued with emergency rule in mind.[48]

Such accounts may have been too willing perhaps to accept the executive claim that 'pressing exigencies' demanded the concentration of power for the sake of the association's protection. Ambivalent towards democracy at the best of times, their authors were perhaps too willing to accept the need for authoritarian solutions. Those of a more constructivist and critical outlook will note that an 'atmosphere of emergency' is easily instrumentalized to galvanize action, cooperation, and compliance, even when the situation at hand is less grave. It can be a governing technique used to head off potential divisions, beyond what is objectively determined.

But thc central point is clear: the diffusion of power across multiple institutions without robust procedures to regulate their interactions establishes the persistent possibility of a drive to centralize power in ad hoc ways in the name of dealing with exceptional situations. As Rossiter put it, 'the more complete the separation of powers in a constitutional system, the more difficult and yet the more necessary will be their fusion in time of crisis'.[49] In the transnational context, where no single institutional agent can claim undisputed authority to act, and where the cooperation of multiple agents is needed to execute measures across legally distinct territories, that fusion is likely to be especially chaotic, and to involve not the emergence of a unitary sovereign but the co-production of emergency rule by multiple executives acting in collaboration.

Emergency Politics in a Minor Key

Let me draw together the chapter's arguments by considering an historical case that speaks to the vulnerabilities described. It illustrates, I suggest, the longstanding potential for this governing mode in the EC/EU context, while stopping

[47] Alexander Hamilton, James Madison, and John Jay, *The Federalist Papers* (Oxford: Oxford University Press, 2008), pp. 99–100.

[48] Hamilton cites the fact that the Romans resorted regularly to a dictator as support for the idea of a strong president (*Federalist Papers*, p. 345): 'Decision, activity, secrecy and dispatch will generally characterise the proceedings of one man in a much more eminent degree than the proceedings of any greater number; and in proportion as the number is increased, these qualities will be diminished.' See also Chapter 8.

[49] Rossiter (*Constitutional Dictatorship*), p. 289. And: 'federalism and local self-government, be they important as in the United States or Switzerland, or rather less significant as in France, present distinct barriers to effective crisis action, and in a great emergency they will be broken down. Constitutional dictatorship must mean a concentration of power, no matter how it is divided.'

short in important ways of what has been witnessed more recently in the 2010s. At its centre is the story of how the European Council came to be established in 1974, a reconfiguration and informalization of executive power aimed at protecting the economic core of European integration in circumstances deemed to threaten it.

The early 1970s were marked, on the one hand, by the renewed desire of leading European governments to push forward the common market programme after a period of uncertainty. The Empty Chair Crisis of the mid-1960s had shown that pursuit of the common policies initiated by the Treaty of Rome could be derailed from within by the reluctance of a member-state to cede powers. As the French Presidency passed from de Gaulle to Pompidou in 1969, many saw an opportunity to renew the integration project by capitalizing on a more favourable outlook in Paris. Included in these calculations was the prospect of extending the market programme to a project of monetary union.[50]

The same period was, on the other hand, characterized by international developments that soon cast doubt on such hopes. The end of the Bretton Woods system (1971) was widely felt to jeopardize existing achievements and further progress. The Commission warned that subsequent US policy would halve the EC's rate of export growth,[51] while fears of monetary instability, inflation, and the disruption of cross-border trade were widespread.[52] The oil shock of autumn 1973 compounded these concerns. Not only did it feed a wider economic recession lasting several years, as well as an international banking crisis in the summer of 1974, but it highlighted the possibility of a breakdown in European policy-making. As individual member-states sought to strike deals unilaterally with oil-producing countries, the capacity to maintain a united front, and to convince external powers like the US that such unity could be relied on, was profoundly in question.[53]

This was the context in which Jean Monnet, re-entering the European stage, sought to convince the major European powers of the need for a 'Provisional

[50] Anthony F. Imbrogno, 'The Founding of the European Council: Economic Reform and the Mechanism of Continuous Negotiation', *Journal of European Integration* 38 (6) (2016), p. 725; Emmanuel Mourlon-Druol, 'Filling the EEC Leadership Vacuum? The Creation of the European Council in 1974', *Cold War History* 10 (3) (2010).

[51] Imbrogno ('Founding...'), p. 726, citing AEI, European Community No. 10, 1971, 3.

[52] Emmanuel Mourlon-Druol, 'Steering Europe: Explaining the Rise of the European Council, 1975–1986', *Contemporary European History* 25 (3) (2016), p. 413: 'Faced with multiple international crises, EEC leaders realised that their coordination at EEC level was poor and decided to meet on a regular and frequent basis in order to try to coordinate their response to those international challenges. Among the various problems facing the EEC from the mid-1970s, macroeconomic issues ranked very high, including European monetary instability, inflation, the development of the common/single market and international trade. Since these policy issues had implications beyond the traditional remit of individual ministries, the EEC needed a greater degree of coordination at heads of government level, if not some kind of collective leadership, in order to try to tackle these problems. Even in the absence of concrete results, the ability to present a united EEC front at the international level was something much sought after by heads of government. The European Council offered the ideal institutional setting in which these issues could be discussed.'

[53] Simon Bulmer and Wolfgang Wessels, *The European Council. Decision-making in European Politics* (London: Macmillan, 1987), p. 33.

European Government' that could protect the achievements of European integration under crisis conditions. As he argued, 'international changes that are crucial for the future of Europe are taking place; and the future of every European country risks being compromised for generations to come, unless those countries face the changes with a clear view of the objectives they share.... That is why... the countries which form the European Community... have decided to set up a Provisional European Government.'[54] As he would later reflect, what was required was 'to return again to the springs of power, first to complete the economic union which had too long been lacking in vigour, and then to seek the forms of a closer and deeper union—federal or confederal, I could not say which. That was the point of the Provisional European Government...'[55]

Rather than for the creation of a permanent new institution, the proposal was for a flexible interim arrangement which would strengthen the leadership role of Europe's executives and thereby secure and consolidate the achievements of economic integration.[56] The existing institutional structures of the EC had shown themselves incapable of delivering this agenda, their procedures too ponderous and dependent on consensus: 'the [EC] institutions have been indispensable, and they still are, but, today, they do not meet the need to move quickly and to decide'.[57] Monnet's efforts to pioneer a new institutional format found favour with the governments of France, Germany and Britain, first under Pompidou, Brandt, and Heath, then under Giscard d'Estaing, Schmidt, and Wilson,[58] and culminated in December 1974 in a communiqué announcing the creation of a new forum of decision-making to be known as the European Council.[59]

This summit-based model of decision-making was a significant departure from the Community's existing norms. Such an arrangement was not foreseen in the Treaty of Rome, nor was it recognized as part of the 'Community method' employed over the intervening years. Few summits of EC heads of state had been held up to this point. Recourse to this decision-making mode was an act of deviation, albeit in the name of better upholding the Treaty's goals of economic coordination. It should be understood as an effort by executive agents to escape

[54] Monnet's memo ('Constitution and Action of a Provisional European Government') was drafted in August 1973 and promoted with European heads of state in autumn 1973: Monnet (*Memoirs*), p. 503.

[55] Monnet (*Memoirs*), p. 509.

[56] See Pascal Fontaine, 'Jean Monnet's Role in the Birth of the European Council' (Sept. 1979). https://www.cvce.eu/content/publication/2006/12/7/ad29595e-0b0a-49b7-ae65-c41c54344a41/publishable_en.pdf

[57] Cit. in Fontaine ('Monnet's role'). Cf. Imbrogno ('Founding...'), pp. 726ff. on Monnet's impression that Community decision-making had become 'too bureaucratic'; cf. Mourlon-Druol ('Leadership Vacuum'), p. 319.

[58] Though the British objected to the name ('provisional European government'): Monnet (*Memoires*), p. 506.

[59] See Final communiqué of the meeting of Heads of Government of the Community, Paris, 9–10 Dec. 1974, in *Bulletin of the European Communities*, 12 (1974).

the constraints posed by existing norms and to concentrate their authority. It was about enabling more decisive action in response to urgent challenges, and creating a platform on which decisiveness could be communicated both to a wider public and to other global powers.[60]

A crucial feature of the new format known as the European Council was the flexibility and informality of its arrangements. Monnet's plans for a provisional European government had emphasized that its meetings would be personal in nature, without the presence of support teams, and its deliberations largely secret.[61] The distinctiveness of the institution was later recognized in these terms from within, in words that anticipate contemporary descriptions of EU politics.[62] As the British briefed Margaret Thatcher before her first meeting in June 1979, 'the European Council is not an institution set up by the Treaties and its formal status is nowhere very clearly defined—partly in order to preserve its informality of procedures'.[63] This ambiguity of status was long maintained: only with the Single European Act (1986/7) was the European Council mentioned in the EU treaties, and only with the Lisbon Treaty (2007/9) was it identified as an EU institution.

Looking back from the present, it is tempting to see the formation of the European Council as part of the logical development of the EU's institutional structure. Its eventual incorporation into the Union framework gives it with hindsight the look of something procedural and complementary to the whole—another step in the emergence of the 'Council system'.[64] Its role in advancing the EMU programme in the late 1970s may suggest its indispensability to the functional goals of economic integration.[65] But to view things this way is to normalize them retrospectively. It overlooks the extent to which what became known as the European Council was a new departure at the time of its introduction, one of

[60] Monnet to Heath (as reported in Monnet (*Memoires*), p. 505): 'We must give public opinion the feeling that European affairs are being decided: today, people have the impression that they're merely being discussed.' Heath agreed. Cf. Mourlon-Druol ("Steering"), pp. 425–6.

[61] Monnet (*Memoires*), p. 504: 'Its deliberations will be restricted to its members, who will keep them secret. The Provisional European Government will have no administrative staff. It will appoint a Secretary, who will minute only its decisions.'

[62] See Djisselbloem's description of the Eurogroup as cited in Chapter 1.

[63] Cit. in Mourlon-Druol ('Steering'). For similar observations in scholarship, see e.g. Wolfgang Wessels, *The European Council* (Basingstoke: Palgrave Macmillan, 2015), p. 42: 'The vague role definition of the European Council enabled them [Heads of State and Government] to exercise powers in a multi-institutional arrangement without being subjected to too many procedural constraints.' See also Emmanuel Mourlon-Druol, 'The Victory of the Intergovernmental Method? The Emergence of the European Council in the Community's Institutional Set-up (1974–77)', in Daniela Preda, Daniele Pasquinucci (eds), *The Road Europe Travelled Along: The Evolution of the EEC/EU Institutions and Policies* (Brussels: Peter Lang, 2010), pp. 36ff.

[64] For an account of this kind: Jeffrey Lewis, 'The Council of the European Union and the European Council', in José M. Magone (ed.), *Routledge Handbook of European Politics* (Abingdon: Routledge, 2015).

[65] Often cited as an achievement of the European Council is the initiation of the European Monetary System (1978), propelled by three Council summits in 1978 in Copenhagen, Bremen and Brussels.

uncertain duration ('provisional' in Monnet's terms), and rationalized by appeal to exceptional transnational circumstances jeopardizing the achievements and prospects of European integration. It was an extension of executive discretion, clinched under the sign of emergency.[66]

Some qualifications are in order. Instructive as this case is in revealing the potential for such moves, in a number of respects it stops short of what one would witness in the 2010s. It was a milder form of exceptionalism in a number of ways. The European Council's creation did not result immediately in the introduction of substantive new policies, or new measures to reinforce existing ones. In its first three years its effects were limited. The intent to act was signalled by soliciting a report from the Belgian PM Tindemans on how to proceed with economic coordination and institutional change, but its impact was slight in the short term.[67] This slower pace of change relative to the events of the Euro crisis stemmed in part from the different challenges at hand. In contrast to more recent patterns examined in the next chapter, executive power was not confronted with a world of hard deadlines emanating from the socio-economic sphere: the timescale for action was more elastic.[68] Equally, there was a shared desire to maintain the principle of consensus amongst equal states. The informality of the European Council was used at first more to discuss common challenges than to impose the will of certain agents on others. And to the extent that public opinion was considered, it was as an audience to be impressed rather than intimidated. Heads of state, still broadly convinced of their capacity to lead and represent their populations, aimed to deliver on a programme for which public consent was assumed. Generating acceptance for one known to be contentious was not yet the order of the day.

What the making of the European Council in 1974 showed then was not that all ingredients of transnational emergency rule as we see it today were in place. Its significance was rather as an illustration of potential: of how and why these practices might emerge, and how the structure of the transnational sphere might foster them.

Conclusion

European integration has long been premised on a dichotomy of order and disorder. Amongst its key rationales has been to move from a world of conflict

[66] Similar impulses would find outlet more widely some months later with the emergence of the G6 and G7: Emmanuel Mourlon-Druol, 'Managing from the Top: Globalisation and the Rise of Regular Summitry, Mid-1970s–Early 1980s', *Diplomacy and Statecraft* 23 (4) (2012), pp. 679–703.

[67] See Tindemans' 'Report on the European Union', requested at the inaugural Council meeting in Paris in December 1974, published 29 December 1975, and put to the Council on 2 April 1976.

[68] On the slow-motion character of crisis management in this period: Hartmut Kaelble, 'Eine beispiellose Krise? Die Krise der europäischen Integration seit 2010 im historischen Vergleich', in M. Heidenreich (ed.), *Krise der europäischen Vergesellschaftung?* (Wiesbaden: Springer, 2014).

between states to one of coordinated policy-making, centred on economic activity but with implications beyond this. Insofar as the economic argument for European integration has often been paired with a security argument to do with peaceful relations between states, the intellectual foundations of the EU have persistently referenced the anarchic alternative. Europe as a political association has been conceived in contrast to something volatile and uncertain.

It is not difficult to see that this idea of the transnational as an escape from a fractious nation-state system combines easily with the politics of emergency. A dichotomy of order and disorder invites challenges to the association to be seen as harbingers of a return to the anarchic condition. As a constructed arrangement, and recent in its construction, the association is easily seen as a fragile achievement, to be defended vigorously against threats always liable to be existential. In the EU this has translated into the protected status of economic policies associated with the common-market programme. Construed as the foundation of cross-border cooperation, the structures of the EU have been built to advance these ends and tailored to their pursuit.

The significance of procedural forms and the norms they give expression to has long been formally acknowledged in this process. The workings of the EU institutions, and their implications for member-state systems, have been circumscribed by conventions of practice and statements of principle. By advocates and architects of European integration, the importance of political and legal norms has generally been recognized: Monnet himself was not deaf to such considerations. But these constraints have been weakly undergirded by sanctions for their breach, and in an association defined by its functional focus they have been secondary in status to an expanding set of policy ends. Moreover, given the challenge of pursuing these ends in a non-hierarchical system and adjusting them to changed conditions, emergency politics carries appeal as a way to galvanize action.

That emergency rule is a persistent possibility does not mean it is persistently invoked. In the history of European integration one finds traces of improvisation rationalized in these terms—of emergency politics in a minor key. But much of this history has been a story of incrementalism rather than dramatic usurpation. The so-called Community Method, in which decisions would be made by Council representatives on the basis of Commission proposals that sought to enact treaty provisions, was the institutional expression of this. The politics of emergency would remain in the shadows as long as this ethos of coordination was the dominant one. Yet the nature of the transnational setting has meant a susceptibility to exceptionalism was there from the beginning.

We have said little so far of the role of mass publics in countering such practices. Contemporary analyses of emergency rule, attuned to the dangers of executive agents over-reaching the demands of strict necessity for more general aims of self-empowerment, often highlight public oversight as a key factor in limiting recourse to it. In this perspective, it is the vigilance of the public that is the best check on emergency rule, and on which even well-developed constitutions and judiciaries must ultimately rely.[69] Where the habits of scrutiny are well established, executive powers must reckon with informal and spontaneous forms of opposition. Where public scrutiny is weak, claims to necessity can be more confidently deployed, since the capacity to cast doubt on them depends on some familiarity with the system. Arguably the transnational sphere is one where public vigilance has tended to be weak and less able to exert a disciplining effect.[70]

Yet equally, where scrutiny is weak, executive discretion may be possible without the drama of accelerated decision-making and appeals to exceptional circumstances. Emergency rule implies one or several audiences in need of convincing: it implies a problem of authority that needs to be tackled. To the extent that transnational politics escapes public notice, or citizens are willing to accept the representative claims made on their behalf by executives, emergency rule is likely to be marginal, certainly as a phenomenon of mass politics. So long as governing agents had little reason to expect populations to react negatively to their actions, they could proceed in more low-key fashion.[71] Should conditions change, however, such that executives were to become more disposed to make abrupt departures from the existing order, while at the same time they could expect more resistance when doing so, the temptations of emergency politics would grow.

This account of the EU's structural vulnerabilities gives us a reading at odds with some conventional defences of it. What are sometimes regarded as its virtues may better be understood as features that expose it to a more authoritarian style of rule. To the extent that it is a purposive association intended to deliver on a delimited set of tasks, it is not necessarily an association constrained to a circumscribed field of activity: that same delimitation of tasks may have the opposite consequence of inviting far-reaching measures in the service of those purposes. While many have regarded the EU as a community founded on law, the orientation of its rule-making to the policies of the common market programme has left some of its basic legal and political norms always a little vulnerable when challenging conditions can be cited. And although the diffusion of power across multiple institutions, and an emphasis on decision-making by consensus, promises

[69] Mark Tushnet, 'Emergencies and the Idea of Constitutionalism', in Tushnet (ed.), *The Constitution in Wartime: Beyond Alarmism and Complacency* (Durham NC: Duke University Press, 2005).

[70] William Scheuerman, '*Die Globalisierung von Carl Schmitt?*', *KJ Kritische Justiz* 50 (2017), pp. 30–7: 'Postnationale Notstandsakteure haben immer noch ein Merkmal, das gewöhnlich mit klassischen Vorstellungen der Exekutive verbunden wird: Die Entscheidungsfindung findet im Geheimen, oder zumindest im Halbdunkel, statt. Dies liegt zum einen daran, dass ihre Transparenz weniger überprüft wird als dies für nationale Exekutiven der Fall ist; zum anderen liegt es schlicht an der Komplexität postnationaler Exekutivgewalt.'

[71] On Europe's 'permissive consensus' before the 1990s: Lisbet Hooghe and Gary Marks, 'A Postfunctionalist Theory of European Integration: From Permissive Consensus to Constraining Dissensus', *British Journal of Political Science* 39 (1) (2009), pp. 1–23.

a political form in which hierarchy can be avoided, at the same time it paves the way for power-political solutions when these methods seem to be failing.

As the next chapter argues, these structural vulnerabilities to emergency rule have been joined in recent decades by tendencies that push strongly in the same direction. Historical changes in the socio-economic domain have made it more tempting to make a diagnosis of extreme conditions and thereby to invoke the logic of emergency rule. The evolution of executive decision-making, notably the rise of an ethos of 'governance' that reinforces the instrumentalist attitude, has made executive powers more inclined to adopt this mode of rule. The vulnerabilities described have come to be exposed more acutely.

3

Towards Emergency Europe

In a celebrated analysis published in 2007, soon after the US invasion of Iraq and the devastation of New Orleans by Hurricane Katrina, Naomi Klein described the emergence of a governing method she termed the 'shock doctrine'. Originating in the 1970s and applied repeatedly thereafter, this entailed the strategic use of disasters as a way to promote socio-economic policies consistent with neo-liberalism. Placing the emphasis squarely on the influence of an ideology, Klein explained how the outlook of national governments and transnational institutions such as the IMF and World Bank came to be influenced in this period by the ideas of Milton Friedman and his fellow travellers, and how the politics of emergency became their preferred method of reform. 'For economic shock therapy to be applied without restraint—as it was in Chile in the seventies, China in the late eighties, Russia in the nineties and the U.S. after September 11, 2001—some sort of additional major collective trauma has always been required, one that either temporarily suspended democratic practices or blocked them entirely.'[1] Policies that might otherwise elicit resistance could thereby be instituted more easily. 'I call these orchestrated raids on the public sphere in the wake of catastrophic events, combined with the treatment of disasters as exciting opportunities, "disaster capitalism".'[2]

One way to read the events of the Euro crisis is as the application of the shock doctrine to the EU. In the context of economic crisis, Europe's executive elites were able to take advantage of generalized disorientation to introduce policies embedding neoliberal preferences. Capitalizing on the structural vulnerabilities of the transnational sphere, they used the techniques of emergency rule to pursue a defined economic and political project. Recent developments in the EU extend an existing template to a part of the world previously largely spared.[3]

Supporting such a reading, the prescriptions of the Troika in the years 2010–15 matched fairly closely the orthodoxies of neoliberal thought as these emerged in the 1990s. The growing influence of Chicago School economics in the Commission and ECB had already been widely observed, and seemed to find ample outlet in the emergency decisions of the period. There was also an undercurrent of

[1] Naomi Klein, *The Shock Doctrine: The Rise of Disaster Capitalism* (London: Penguin, 2007), p. 11.

[2] Klein (*Shock Doctrine*), p. 6. For some updating of these themes: Naomi Klein, (2017), *No is Not Enough: Defeating the New Shock Politics* (London: Allen Lane, 2017).

[3] Not all of Europe had been spared: see EU-IMF cooperation in non-Euro states Hungary, Latvia, and Romania 2008–9, as well as the broader 'transition' process in 1990s Eastern Europe.

Politics of Last Resort: Governing by Emergency in the European Union. Jonathan White, Oxford University Press (2020). © Jonathan White.
DOI: 10.1093/oso/9780198791720.001.0001

opportunism in these actions, with executives invoking crisis conditions as reason to advance measures long sought.[4] Of course, the world Klein described was one in which neoliberals could present themselves as the bearers of solutions in part because the shocks faced were widely seen as exogenous to neoliberal capitalism; in 2010s Europe, they would have to contend with the suspicion that their own policies and omissions were implicated in the problems at hand. But arguably they were rather successful in presenting their solutions as needed. A convincing case can be made that emergency measures sought not just to stabilize the economy but to do so in a way that would shrink the welfare state and advance privatization and competition, in the Mediterranean South and across the EU more widely.[5] Thinkers associated with neoliberalism had long seen in European integration an opportunity for the advancement of their cause: arguably the events of the Euro crisis confirmed their intuitions.

A relatively minor corrective is to note how the drivers of EU emergency politics extend beyond the influence of neoliberalism alone. One sees the involvement of agents whose motivations are complex and sometimes hard to reconcile with Chicago School economics.[6] In some institutions, notably the IMF, strands of Keynesianism were arguably on the rise throughout the period in question.[7] Of interest then is how those not necessarily committed to the neoliberal project nonetheless came to take up a leading role. Further, the practices of emergency rule have been extended to policy domains such as migration that are connected only indirectly to neoliberal concerns. The practices of emergency seem compatible with a diversity of ideological projects.

But a more interesting refinement of Klein's argument—perhaps better: a radicalization—concerns how the EU's politics of emergency plays out in connection with the structural features of a transnational polity. Rather than external technocratic institutions imposing themselves on an isolated elected government from without, for the most part one sees the willing coordination of the two, enacted before the representatives of multiple interdependent states. Emergency politics is in this sense genuinely *transnational*, inspired, as I shall argue, by a shift in the ethos of executive power that permeates both state and supra-state institutions. If shock doctrines find a hearing in this context, it is because they fit with patterns of rule predating any one period of malaise, and capitalize on the networks

[4] E.g. Wolfgang Schäuble (2011), 'We can only achieve a political union if we have a crisis', *New York Times*, 18 November (http://www.nytimes.com/2011/11/19/world/europe/for-wolfgang-schauble-seeing-opportunity-in-europes-crisis.html).

[5] E.g. Costas Lapavitsas, *Crisis in the Eurozone* (London: Verso, 2012). As evidence of intent, see again the leaked letters of ECB head Trichet to Ireland, Spain, and Italy.

[6] On the question of Ordoliberal influence: Hien and Joerges (*Ordo-Liberalism*). On the ECB as ideologically committed to austerity, but the outcome of its interactions with Germany, France, and the IMF as something more varied than neoliberalism: Adam Tooze, *Crashed: How a Decade of Financial Crises Changed the World* (London: Penguin, 2018), pp. 341ff.

[7] Ben Clift, *The IMF and the Politics of Austerity in the Wake of the Global Financial Crisis* (Oxford: Oxford University Press, 2018), p. 4.

of transnational coordination associated with EU membership. Moreover, there is a cumulative dimension to the process. Emergency measures do not just herald new policies but redistribute powers amongst agents in ongoing relations of interdependence. There is a polity-building aspect to proceedings, sometimes involving the creation of new institutions, and often with the potential for further rounds of emergency rule in future.

As this chapter argues, the EU's entanglement with emergency politics should be seen as driven by mounting challenges to the authority of representative institutions. These have been led on the one hand by changes in the societal context, with the rise of non-state agents able to define challenging situations as crisis moments demanding action. Such securitizing agents apply pressure directly to executive authorities, and see it magnified by the plurality of executives in the EU sphere and the need of each to signal to peers their readiness to uphold the common policies to which they are committed. These aspects are examined in the first section.

In parallel, changes internal to executive power have created a decision-making environment receptive to these emergency claims. Political parties that define themselves by ideologies of principle, and that might therefore resist securitizing pressures emanating from the private sphere, have tended to be marginalized by those embracing an ethos of managerialism and problem-solving.[8] This reinforces the instrumentalist orientation observed in the previous chapter, increasing the detachment of decision-making elites from mass publics, and encouraging resort to the techniques of emergency rule so as to cope with their diminishing authority. More power is transferred to unelected institutions, where the same managerial ethos is especially pronounced, and where the provision of 'security' in all its forms gels with an agenda of delivering outputs presumed favourable to all. These patterns are examined in the the second section. As becomes clear, executive institutions seek in the politics of emergency an authority of last resort, one that can address some of the concerns of public opinion while responding to demands that are generated elsewhere.

Central to the politics of the Euro crisis, how far are these dynamics visible in the EU's handling of other challenges? In the chapter's third section I take on this question by considering European migration policy in the years after 2015. The suggestion is that, notwithstanding some major differences, much the same template of emergency rule was present here too, driven by similar factors. While the Euro crisis remains the book's principal reference-point, the comparison suggests the concepts used to analyse the one case are transferable to others.

The structural vulnerabilities discussed in the previous chapter mean the EU was never well protected from emergency politics. What this chapter identifies are

[8] Mark Blyth and Richard Katz, 'From Catch-all Politics to Cartelisation: The Political Economy of the Cartel Party', *West European Politics* 28 (1) (2005), pp. 33–60.

the dynamics by which this long-standing susceptibility comes to be acted on. Here our focus extends to factors not necessarily peculiar to the EU realm, but to which its vulnerabilities leave it especially exposed, and by which it is shaped in distinctive ways. The EU gives advanced expression to a wider remaking of administrative power that reinforces discretionary over procedural forms of rule.[9] While neoliberal economics has been part of the process, it stands as one dimension of a more general challenge to the authority of representative institutions to which emergency rule responds.

The Rise of Extreme Conditions?

Emergency rule has always been theorized in conjunction with challenging circumstances. Part of its logic involves appeal to the licence of exceptional conditions: wars, natural disasters, or economic upheaval. Caution tends to be warranted concerning how far such circumstances are really exceptional: long-term factors are always in play. The structure of the polity may also shape the extent to which challenging situations escalate into something graver. But these points accepted, one may reasonably ask whether tendencies towards emergency rule reflect developments in the extra-institutional sphere. How far can we attribute the transnational politics of emergency to the rise of extreme conditions inviting a response of this kind?[10]

Macro-historical tendencies of various kinds might seem to point in this direction. I shall highlight just a few, in policy domains relevant to the EU. From the perspective I am advocating, however, more important than such material drivers of emergency rule are the interpretations given by powerful agents. Rather than simply 'extreme conditions' as such, we are dealing with the capacity of interpreting agents to credibly portray situations as extreme and in need of urgent address. Their influence is observable on politics at all levels, but carries especially far in the transnational setting on account of the structural features outlined in the last chapter.

A number of strands of contemporary sociology argue that politics today must contend with a social world that is increasingly volatile. Such arguments come in various forms, ranging from diagnoses of *social acceleration*, to diagnoses of a new epoch of *risk*, to diagnoses emphasizing the rise of financial capitalism.[11] Changes

[9] Honig (*Emergency Politics*), pp. 66ff.

[10] Using this terminology: Thomas Eger, Stefan Oeter, and Stefan Voigt, *International Law and the Rule of Law under Extreme Conditions: An Economic Perspective. Contributions to the XIVth Travemünde Symposium on the Economic Analysis of Law (March 27–29, 2014)* (Tübingen: Mohr Siebeck, 2017).

[11] On theses of 'social acceleration': Scheuerman (*Liberal Democracy*); Hartmut Rosa, *Social Acceleration: A New Theory of Modernity* (New York: Columbia University Press, 2013). On risk

in the socio-economic sphere are a focus for many, and carry important implications for transnational politics. As many observe, changes in the global economy since the 1980s have contributed to a more uncertain and fast-shifting set of conditions. Jessop speaks of 'the state's loss of time sovereignty'.[12] Forced to reckon with such nebulous phenomena as investor confidence in the digital age, things that fluctuate unpredictably and apparently resist responses based on stable rules and regularized procedures, decision-makers are said to be faced with a qualitatively new set of conditions for governing.

In accounts of this kind, it is not just that the socio-economic context of decision-making is fast-changing, but that the stakes of public policy are especially high under conditions of interdependence. For most OECD countries, the decades since the mid-1970s have been a time of rising public and private debt.[13] Moreover, the inability of any sizeable entity—state, financial institution, or aggregate of small borrowers—to service its debt will hit other indebted agents hard too. The interconnectedness of the global banking system provides a mechanism by which local difficulties are rapidly transmitted across territorial and functional space.[14] Extreme conditions at a systemic level become a perennial prospect.

Confronted with these challenges, decision-makers may find two options tempting. On the one hand they may abdicate responsibility for policy in domains where these conditions of volatility obtain. Privatization and welfare-state retrenchment are two of the familiar outcomes. On the other—particularly when such an approach becomes untenable—they may be tempted to reassert themselves, not through the conventional, slow channels of political procedure but by means of 'fast policy'[15] designed to respond quickly to fast-moving and unpredictable events. Such interventions typically aim to deal with immediate challenges in their specificity rather than to establish lasting principles. They are consciously exceptional. The overall effect, in one author's words, is that 'liberal democratic lawmaking increasingly consists of vague and open-ended legal resolutions, exceptional and emergency norms delegating vast discretionary power to

society and its implications for governing, see i.a. Ulrich Beck, 'Critical Theory of World Risk Society' *Constellations*, 16 (1) (2009).

[12] Bob Jessop, 'The Spatio-Temporal Dynamics of Globalising Capital and their Impact on State Power and Democracy', in Hartmut Rosa and William Scheuerman (eds), *High-Speed Society* (University Park: Pennsylvania State University Press, 2009), p. 154.

[13] Public debt rose across OECD countries from the mid-1970s, from c. 40% of GDP to >90% by 2010 (Wolfgang Streeck, 'The Politics of Public Debt: Neoliberalism, Capitalist Development and the Restructuring of the State', *German Economic Review* 15 (1) (2014), p. 145), while private debt rose throughout this period.

[14] Tooze (*Crashed*), pp. 9ff.

[15] Jessop ('Spatio-Temporal Dynamics'); as deployed transnationally: Arjen Boin, Magnus Ekengren, and Mark Rhinard, *The European Union as Crisis Manager: Patterns and Prospects* (Cambridge: Cambridge University Press, 2013).

the executive and various administrative bodies, and poorly crafted statutes possessing a limited half-life'.[16]

Tendencies towards volatility in the wake of global socio-economic change since the 1980s carry powerful implications when they meet an idea of transnational politics rooted in the common-market programme. As we have seen, a purposive association elevates certain goals as constitutive of its reason to exist. To the extent that environmental conditions are relatively stable, these purposes may be pursued by the incremental adjustment of existing mechanisms and with sensitivity for politico-legal norms. But this logic of fidelity depends on the past being minimally 'useable'. Where situations arise for which the inherited structures seem ill-suited, there is the temptation to cut loose. For the sake of preserving the ends of the association in the face of apparent threats, far-reaching departures from existing commitments may be sought. And because the 'fast policy' that responds to such challenges may often be inadequate for tackling their causes, its legacy will often be an uneasy cross-border interdependence that generates further conditions of volatility.

In this way the EU becomes increasingly susceptible to emergency rule, as the long-term factors discussed in the last chapter combine with more immediate pressures. This, at least, would be the tempting conclusion to draw—but I want to tweak the argument's focus. Whether 'extreme conditions' are on the rise in some absolute sense is something we may leave to the critical economists and sociologists, since more directly relevant to the patterns of rule examined is how objective conditions come to be determined. Material tendencies do not act directly on executives. What I want to emphasize is the rise of intermediary agents, one step removed from the halls of decision-making, with the authority to interpret socio-economic conditions, to make sense of volatility, and to specify the responses it demands and when. What can be discerned, I suggest, is the rise of authoritative agents only too inclined to cite conditions of crisis, fostering an atmosphere conducive to emergency rule.

The economic domain provides some of the most important examples. In an age of generalized indebtedness, *confidence* becomes the sought-after quality, for which authoritative assessments are sought. The rise of public and private debt corresponded to the rise of bond markets and ratings agencies, as those claiming the authority to quantify it, to establish the reliability of debtors and their fiscal capacities, to shape the expectations of investors, to set and monitor deadlines, and to establish the point at which reasonable risk tips over into a situation demanding action by investors and authorities. Especially with the growing importance of creditworthiness after the demise of the Bretton Woods system, such agents increasingly became the measure of socio-economic conditions. By

[16] Scheuerman (*Liberal Democracy*), p. 56.

the turn of the millennium, these *spokespeople of the markets* had, provisionally at least, displaced states as the preeminent sources of data, measurement, and analysis.[17] Their market interpretations could now shape what should be seen by decision-makers as an urgent economic situation in need of address. In turn, the secondary interpretations of market analysts, consultants, insurance companies, banks, investors, financial elites, think-tanks, and accounting firms could shape what counts as the necessary response on the part of authorities.[18]

Such intermediary figures engaged in the determination of socio-economic conditions, and the identification of when they are 'extreme', become crucial in establishing the circumstances of emergency rule. What they acquire is the power to *securitize* situations: i.e. to identify when certain problems are to be elevated to the status of fundamental threats to the existing order, and to identify the prescriptions that should follow. Objective conditions do not interpret themselves: emergency situations need identification as such, in particular because what are taken to be the relevant threats are often things yet to happen (e.g. currency runs). Taking their measure depends on extrapolating from the present to the near future, which foregrounds those with the authority to interpret uncertain conditions.

As private providers of information, many such agents exist in a competitive relation and have reason to emphasize the severity of the circumstances they respond to. Those who speak in the most urgent tones, and who time their interventions for dramatic effect, may well be the most likely to be heard. When public authorities are dependent for information-gathering and evaluation on private entities of this kind, they are dependent on those inclined to magnify the stakes of the moment. Moreover, they are dependent on those likely to prioritize certain considerations over others in their assessment of what constitutes an emergency. The prospect of loan defaults is likely to eclipse other concerns to do with everyday standards of living. Arguably one reason why executives become so averse to the prospect of defaults is because the system of credit rating is built on the idea that such outcomes are amongst the worst that can happen. In this way there is a filtering effect on the determination of what matters.

[17] Benjamin Lemoine, 'Measuring and Restructuring the State: Debt Metrics and the Control of Present and Future Political Order', in Desmond King and Patrick Le Gales, *Reconfiguring European States in Crisis* (Oxford: Oxford University Press, 2017); cf. Bartholomew Paudyn, 'Credit Rating Agencies and the Sovereign Debt Crisis: Performing the Politics of Creditworthiness through Risk and Uncertainty', *Review of International Political Economy* 20 (4) (2013), pp. 788–818.

[18] On the increasing importance of evaluations of trustworthiness as debt rises: Wolfgang Streeck, *Gekaufte Zeit. Die vertagte Krise des demokratischen Kapitalismus* (Berlin: Suhrkamp, 2013), p. 155; on the nexus of financial and executive power: Vogl (*Ascendancy of Finance*). On instances when the announcements of ratings agencies escalated Eurozone uncertainty in 2010: Tooze (*Crashed*), pp. 338, 385.

These mechanisms were at work in spring 2010 in framing the situation in Greece and other troubled states as one of high emergency.[19] By offering a definition of the situation as a 'sovereign debt crisis' centred on the government's finances and its capacity to repay loans, by offering credible measures of the situation's gravity, deadlines for action, and a reference-point for evaluating the policy initiatives that 'work', these spokespeople of the market shaped the nature and pace of the political response, and put obstacles in the path of those who might seek a different path.[20] The emergence of 'the spread' as a yardstick for evaluating and steering decision-making in the Eurozone is a salient example. Referring to the yield premium that investors would demand to hold government bonds considered riskier than Germany's, it became influential in the choreography of crisis measures. The spread defined the emergency as one of the credibility of state finances, quantified the unfolding sense of urgency, and, being based on a transnational comparison, enabled pressure to be localized on particular governments while ensuring that other governments and the wider socio-economic order were external to the problem-definition. Such tools of evaluation became the mechanisms by which private agents came to shape crisis measures and their public narration.[21] One may further note that, insofar as emergency politics deals not just with threats already manifest but those said to be emerging or imminent, and thus requires the interpretation of an uncertain future, the focus of these agents on *forecasting* gives extra capacity to influence.[22]

The field of securitization studies, emerging in the 1990s with the Copenhagen School, has traditionally associated the ability to frame a situation as an emergency with the organs of state. Increasingly this looks incomplete, however. Non-state figures must now be recognized both as securitizing agents in their own right, and as an audience that can validate the securitizing moves of executives.[23] Whether or not they actively seek to incite emergency measures will vary. Profit motives may sometimes encourage this, e.g. in the form of currency speculation, but stability may often be as attractive, certainly once favourable changes have been achieved. Rather than that these intermediaries consciously seek to manufacture extreme conditions, what matters is that they

[19] For an instructive overview of the authority of ratings agencies at the outbreak of the Euro crisis and their influence in shaping the response: Richard Milne, 'Credit Rating Agencies Place ECB in a Bind', *Financial Times*, 5 July 2011: https://www.ft.com/content/c09f18b6-a71a-11e0-a808-00144feabdc0

[20] Andreas Langenohl, 'Securities Markets and Political Securitization: The Case of the Sovereign Debt Crisis in the Eurozone', *Security Dialogue* 48 (2) (2017), pp. 131–48; cf. Mark Blyth, *Austerity: The History of a Dangerous Idea* (Oxford: Oxford University Press, 2013).

[21] On 'the spread' in Italian emergency politics in the 2010s: https://www.ilpost.it/2015/03/07/storia-spread/

[22] Jens Beckert, *Imagined Futures: Fictional Expectations and Capitalist Dynamics* (Cambridge MA: Harvard University Press, 2016), pp. 122–3, pp. 218ff.

[23] See also Langenohl ('Securities'). On the difference between agents who can securitize and those who can act decisively in light of this: Thierry Balzacq, Sarah Léonard, and Jan Ruzicka, '"Securitization" Revisited: Theory and Cases', *International Relations* 30 (4) (2016), p. 501.

are permanently attentive to their possibility. They are consistently monitoring evidence for signs of malaise, are willing to talk up the significance of the moment, and have the authority to imprint their interpretations on others.

Clearly these are not developments unique to Europe. To highlight such factors is to highlight long-term accruals in socio-economic power that go beyond any one region. But such developments are nonetheless of special relevance to a transnational order such as the EU. Because their monitoring covers multiple jurisdictions and serves economic interests with no particular relation to any one country, market interpreters are professionally inclined to take the transnational view and to cast extreme conditions in these terms. They are structurally disposed to 'macro-securitization', i.e. to the identification of threats to interests and policy regimes that are cross-border in range.[24] Often the measures of order and disorder they pioneer are based on cross-country comparisons, putting national executives in a relation of competition and inviting the aerial view.

It is the *transnational* politics of emergency that they foster, and the structure of the transnational realm that amplifies their influence by making their assessments the basis for peer monitoring. The plurality of executive agents in the EU sphere, and the concern of each with the maintenance of common policies, means warnings of disorder are likely to be taken seriously, invoked both as reasons for action and to galvanize others. With no single executive able to guarantee order, and at least partly dependent on decisions taken elsewhere, each is disposed to be watchful of its peers and alert to disquieting information. Executives may recycle the warnings offered to them by market agents and re-direct them at each other and domestic institutions, hoping thereby to impress on them the need for action. At different moments these executives are both the addressees and the makers of emergency claims. The structure of the transnational polity invites a multi-dimensional and processual set of dynamics, rather different from conventional accounts of securitization.[25]

I have dwelt on the economic domain given its capacity to influence others and its centrality to the EU, but the increasing significance of non-state agents extends beyond this. Media outlets have come to enjoy a similar capacity to determine extreme conditions. Whether in relation to natural disasters such as earthquakes, or more obviously human-made situations involving large-scale migration, they are well placed to make emergency claims that frame circumstances as exceptional and in need of response. A mass media concerned with dwindling advertising revenue has reason to amplify such claims to sell the news of the day. Here too one

[24] On 'macrosecuritization': Barry Buzan and Ole Wæver, 'Macrosecuritization and Security Constellations: Reconsidering Scale in Securitization Theory', *Review of International Studies*, 35 (2) (2009), pp. 253–76.

[25] On the limits of securitization accounts resting on speech-act theory and the need to focus on multilateral, processual interactions: Adam Côté, 'Agents without Agency: Assessing the Role of the Audience in Securitisation Theory', *Security Dialogue* 47 (6) (2016), pp. 541–58.

sees a power of securitization. Newspaper websites cultivate a sense of the exceptional situation whenever they magnify their font size or devote the top of their homepage to the reporting of a single story.[26] They prime their readers, and those wanting to second-guess them, to feel they are in exceptional times. Insofar as they pursue a transnational audience online, they too have reason to transnationalize the extreme conditions they report.

In sum, socio-economic changes in recent decades have given greater scope for the diagnoses of extreme conditions on which emergency rule rests. There is an abundance of material well suited to crisis talk, touching in particular on the core policies of European integration. But such conditions do not interpret themselves. Crucial is how their existence has been coupled with the emergence of authoritative figures beyond public institutions able to shape the meaning of conditions and their policy implications. As information-gathering and evaluation is outsourced to entities of the private sphere, the conditions ripen for emergency claims. In the 2010s, the warnings of these spokespersons of the market would be the basis on which executive power, selectively but repeatedly, would find reasons to act. And their influence would increase further as the outlook of executive power became increasingly sympathetic to their claims.

Instrumentalism Unbound: The Changing Ethos of Executive Power

As the contexts of decision-making have evolved in recent decades, so the domain of executive power itself has changed. Developments in the former have helped reshape, indeed have permeated, the latter. In particular, one may highlight a shift in the epistemic foundations of decision-making, one that embeds an outlook favourable to emergency rule. If a certain sensibility for means-ends thinking was something identified in the previous chapter as a long-term feature of European transnationalism, what emerges in this later period is its open embrace. One sees the rise of a governing ethos founded on problem-solving and reliant on discourses of emergency to connect with ever more remote and potentially obstructive mass publics. Often associated with the eclipse of state sovereignty, the *governance* paradigm may seem a surprising place to look for impulses towards the politics of emergency. But if we approach the latter as a distributed, co-produced phenomenon, the continuities become clear.[27]

[26] Mai'a Davis Cross and Xinru Ma, 'EU crises and integrational panic: the role of the media', *Journal of European Public Policy* 22 (8) (2015).

[27] For the 'hint of decisionism' in governance: Henrik Enroth, 'Governance: The Art of Governing after Governmentality', *European Journal of Social Theory* 17 (1) (2014), pp. 60–76; Honig (*Emergency Politics*), p. 67.

Underpinning European integration for much of the latter twentieth century, and offsetting the structural vulnerabilities to emergency rule discussed, were a number of agential kinds that have since become less prominent. The centrality of the legal profession was one. Political sociologists have shown how legal credentials were valued both in the EU's judicial and executive branches. A legalist outlook was part of the ethos of decision-making, to the extent one could speak of the 'legalistic colonization of Europe's political and bureaucratic spaces' in the years after Monnet left the scene.[28] Such an outlook formed a braking mechanism on emergency politics. For as long as it shaped the common-market programme, it would temper these impulses—not because legal authorities are necessarily unwilling to exercise discretion, but because they aim to do so within the frame of procedural norms.[29]

The mid-twentieth century ethos of executive power was also shaped by the residual influence of societally rooted political parties. Although state representation in transnational politics has always concentrated decisions in the hands of party leaders, these leaders remained embedded in organizations committed to mediating their activities with a wider public and connecting them to programmatic ideas. They were surrounded by experts chosen mainly for what they could contribute to the party's agenda rather than their credibility with socio-economic powers.[30] Parties in this mould could be expected to insist on some measure of cross-temporal consistency, acting as a constraint on leaders' discretion. When politicians articulated a vision of the EU, there was the expectation it be squared with their party's ideological commitments: Europe had to be conceived as the extension of a political project.

What one saw emerging from the 1980s was a governing ethos different in focus. Commonly referred to as the rise of *governance*,[31] this decision-making mode seeks its legitimation in problem-solving capacity. As Enroth puts it, it amounts to a shift 'from an art of governing premised on producing policy for a society or a population to an art of governing premised on solving problems with

[28] Cf. Antoine Vauchez, 'The Force of a Weak Field: Law and Lawyers in the Government of the European Union', *International Political Sociology* 2 (2) (2008), pp. 128–44; Stephanie Mudge and Antoine Vauchez, 'Building Europe on a Weak Field: Law, Economics, and Scholarly Avatars in Transnational Politics', *American Journal of Sociology* 118 (2) (2012), pp. 463–4.

[29] On integration 'through law' vs 'through crisis': Nicole Scicluna, 'Integration through the Disintegration of Law? The ECB and EU Constitutionalism in the Crisis', *Journal of European Public Policy* 25 (12) (2018).

[30] On the shift in left parties from theoretician-experts to 'transnational, finance-oriented economists': Stephanie Mudge, *Leftism Reinvented: Western Parties from Socialism to Neoliberalism* (Cambridge MA: Harvard University Press, 2018), pp. 6ff., pp. 217ff.

[31] On governance: Claus Offe, 'Governance: An "Empty Signifier"?', *Constellations* 16 (4) (2009); Wendy Brown, *Undoing the Demos: Neoliberalism's Stealth Revolution* (MIT: Zone Books, 2015), p. 127; Vogl (*Ascendancy of Finance*), p. 144: 'governance refers to the integration of private actors and agencies into the public realm, the proliferation of private-public synapses, and the intensification of reciprocities and interdependencies'.

no necessary reference to any kind of society or population.'[32] At its most basic, governance can be seen as a response to the growing problems for public authority described in the previous section. It marks efforts to undergird the credibility of public policy in the eyes of powerful socio-economic agents by reinforcing its insulation from normative considerations. But in addressing those problems, the practices of governance generated new ones, notably to do with the legitimation of policy to mass publics.

One expression of the EU's turn to governance from the 1980s was the increasing prominence of a certain kind of economic expertise in senior executive positions. On the one hand one saw individuals with economic rather than legal qualifications becoming more central, in the European Commission especially.[33] An economic training would become the more fungible resource for aspiring European officials, especially with the introduction of the Euro. To the extent that legal expertise was still prized, it was less as a global outlook and more as a specialist capacity to be deployed for specific instrumental purposes.[34] New management techniques, soft law and 'comitology' came to the fore, along with increasing recourse to quantification as a way to assess progress in the achievement of goals.

Equally, the *kind* of economic expertise that became prominent was one conceding considerable space for executive discretion, as long as it served approved outcomes such as market liberalization, privatization, creditworthiness, and competitiveness. Such shifts in the EU go well beyond its non-majoritarian institutions. The Council too has been affected, as national parties fall increasingly under the sway of finance-oriented economists who, in Stephanie Mudge's words, view themselves as 'interpreters, guardians, and (sometimes) saviours of markets'.[35] Chicago School economics would not displace all elements of the Ordoliberal tradition—indeed, the latter's lingering influence[36] is one reason to resist seeing the EU as simply playing host to a neoliberal shock doctrine. But one saw an emerging outlook more indifferent to politico-legal normative constraints, and inclined to see the EU in terms of a more narrowly defined set of purposes to do with market competition and price stability.[37]

As executive power becomes increasingly beholden to the ethos of governance, the instrumentalist attitude inherent in a purposive association comes to be

[32] Enroth ('Governance'), p. 61.

[33] On the post-1980s marginalization of lawyers and rise of economists: Didier Georgakakis and Jay Rowell, *The Field of Eurocracy: Mapping EU Actors and Professionals* (Basingstoke: Palgrave Macmillan, 203).

[34] Didier Georgakakis, 'À la fois pilote et victime de l'austérité ? Les transformations de l'administration de l'Union européenne sous l'effet de la crise economique et financière', *Revue Française d'Administration publique* 151–2 (2014), p. 819.

[35] Mudge (*Leftism*), p. 7.

[36] Bonefeld (*Strong State*); Thomas Biebricher, 'Ordoliberalism as a Variety of Neoliberalism', in Hien and Joerges (*Ordo-Liberalism*).

[37] Hien and Joerges (*Ordo-Liberalism*), 'Introduction'.

openly embraced and further detached from the countervailing concerns that might moderate it. Weakly delineated institutional processes (and weakly organized mass publics) appear from the governance perspective as some of the distinctive advantages of the transnational realm, giving more freedom for executive manoeuvre. As a managerial outlook bearing the influence of the private sector, in particular of notions of strong leadership in firms, governance carries over a certain indifference to the conventional criteria for the exercise of public authority.[38] The primacy of substantive ends over procedural constraints finds outlet in a preference for administration by target-setting. Initially a governing method associated primarily with central banking, it has become increasingly prominent in other EU areas too.[39] Formal distributions of power, including the distribution of authority between different institutions and the mandates and procedures associated with each, are generally secondary concerns. Whereas legal approaches take seriously the past as a source of legitimation, and party-ideological approaches tend to attach weight to the procedures of politics, the logic of governance points elsewhere. Its focus is on the moving present and how to address efficiently the challenges encountered, using trans- and para-institutional networks. Tendencies towards governance imply increasing willingness to take advantage of the latent possibilities for unconventional action in the transnational setting.

In the governance frame, decision-making is a collaborative problem-solving process between a variety of agents spanning public and private authorities. What comes to be emphasized are ideas of best practice in addressing the problems of the moment and the concerns of relevant stakeholders, often by involving them in decision-making.[40] The problem-solving ethos offers no reason to push back against the agents of securitization of the socio-economic domain: they are more likely to be invited to the decision-making table. In the extreme case their evaluations are built into the design of public policy, as for example when the assessments of ratings agencies are made the basis of central-bank decision-making.[41] Rather than disputed, their claims are thereby invoked—selectively, to be sure—as warrants for action. Governance takes power as it finds it.

In this way executive authority becomes increasingly reactive, organizing itself in response to what it treats as functional demands. The autonomy of political institutions vis-à-vis market agents, including the willingness to set distinct normative priorities, is undercut. This is not to imply executives merely wish to

[38] Brown (*Undoing the Demos*), p. 41; Offe ('Governance'), p. 556.

[39] On governance and target-setting: Gwyn Bevan and Christopher Hood, 'What's Measured Is What Matters: Targets and Gaming in the English Public Health Care System', *Public Administration* 84 (3) (2006), pp. 517–38; as applied to the EU: Lorenzo Codogno, Guillaume Odinet, and Flavio Padrini, 'The Use of Targets in the Lisbon Strategy', *Rivista di Politica Economica* January–March (2009).

[40] Brown (*Undoing the Demos*), p. 71.

[41] On Trichet's increasing use of market evaluations as a governing technique: https://www.ft.com/content/c09f18b6-a71a-11e0-a808-00144feabdc0; cf. Tooze (*Crashed*), p. 334.

do the bidding of market participants, nor indeed that they always agree on what market-conforming policy looks like. As the handling of the Euro crisis would show, there could be disagreements—e.g. between those of mainly Ordoliberal and neoliberal outlook—concerning how best to respond to the economic pressure in play. Ratings agencies conveyed the need to do *something*, and to do it urgently, but there could be disagreement about precisely what. Market agents themselves might feel that certain responses did not adequately address their concerns: austerity policies were not always met with enthusiasm, even when publicly justified as responses to market pressure. And amongst some governing agents there was apparently the intent, visible in the 'rescues' of Greece, exactly to show that investors do *not* dictate political outcomes. There was the desire to show that political authorities can withstand at least some of the pressures to which they are subject by upholding existing loan agreements and ruling out write-downs.[42] But even if this led policy-makers to avoid policies like debt restructuring, their guiding principle nonetheless was to pursue those that would restore market confidence. The governance outlook meant seeking to accommodate these concerns rather than looking for ways to push back on them.

The rise of this ethos paves the way for emergency rule not just by fostering indifference towards politico-legal norms. Worth noting is its difficult relation to public legitimation. One reason economists became increasingly central to decision-making in this period is arguably because governments hoped the authority of economists could help paper some of the cracks in their own. In place of the integrating capacity of political ideologies, economic expertise would come to the fore. Yet this reconstitution of executive power brings a new set of challenges. Refashioning executive power as a problem-solving network rather than as a set of representatives embedded in party-political and legal traditions, it creates the prospect of a widening separation between executives and citizens, of which the former may be well aware.

Successful problem-solving capacity can be one way of bridging this gap—an appeal to technocratic insight of the kind we shall examine in Chapter 5.[43] But problem-solving accomplishments are a difficult way to narrate decision-making, especially when the accomplishments are in question. Should executives come to doubt the hold of output legitimacy, the politics of emergency makes for an authority of last resort, a way to pursue consent for the initiatives undertaken.

Particularly when validated by figures of the private sphere, such a discourse gives citizens reason to accept decisions that they did not participate in forming and may be reluctant to accept. It is a way to depoliticize when other methods

[42] See Sandbu (*Europe's Orphan*), pp. 55ff. on the French government's motivations, also Trichet's.
[43] See also the notion of 'output legitimacy', developed in EU studies by Scharpf (*Governing in Europe*).

give out.[44] As such it can also weaken parliamentary resistance. For those opposition parties that, like their peers in government, struggle to maintain themselves as associations unified around popular ideologies, emergency politics may offer a welcome rationale for avoiding tough decisions, absolving individuals of responsibility and the burdens of adversarialism. By highlighting socio-economic pressures as reasons for action—pressures readily confirmed by the authoritative figures of that domain—governing agents ease the passage of their decisions. Applied to problems plausibly transnational in scope, such an approach promises a way to integrate diverse populations in the face of common threats and build consent. Emergency politics is a way of contending with diminishing public authority in an age of governance.

For Naomi Klein and others, the rise of the neoliberal agenda, and the 'shock doctrine' of some of its adherents, are the central driving forces. Her work usefully shows that supranational institutions such as the IMF and World Bank have been incubators of a disposition to emergency politics.[45] But what is evident in the EU is the role of trans-institutional governance networks in cultivating receptivity to such ideas and offering a platform for their enactment. Such networks have built the habits of coordination while linking them to a defined set of policy goals to be upheld. And while neoliberalism is undoubtedly part of the story, there is no need to seek the origins in one ideology alone. The ethos of governance is the sum of many shifts in the composition and outlook of executive power, whose effect is to redouble the instrumentalist attitude on which the politics of emergency feeds.

Emergency Politics beyond the Euro Crisis

Europe's period of economic emergency in the 2010s marks the interaction then of several tendencies. There are, first, the EU's long-standing structural vulnerabilities as described in the previous chapter. Built around a set of policies to do with the common market programme, its procedural arrangements have been secondary to the pursuit of its defining ends. Institutionally, it has developed as a complex, divided, and potentially cumbersome arrangement given the reliance on consensus across multiple sites. The diffusion of power creates incentive to re-concentrate it informally in the hands of the few when swift action is sought, all the more because constitutional relations between the parts are weakly codified and underpinned.

[44] White ('Emergency Europe', p. 307). For a similar reading of welfare-state transformation: Zygmunt Bauman, 'From Social State to Security State', in Bauman, *Europe: An Unfinished Adventure* (Cambridge: Polity, 2004).

[45] Klein (*Shock Doctrine*), pp. 162ff.

These long-standing vulnerabilities have become more significant as the external pressures for decisive interventions increase. While 'extreme conditions' are partly the function of deep-seated changes in the socio-economic sphere, more directly they reflect the growing capacity of non-state agents to frame conditions as extreme—to securitize them such that they demand a response. Such moves grant credibility to emergency rule and filter the ends to which it is put. The picture is completed by a shift within the outlook of executive power itself. As the problem-solving ethos of governance has become pronounced in both national and supranational, political, and technocratic institutions, instrumentalism has come further to the fore. Decision-making falls under the sway of those comfortable in overriding norms on consequentialist grounds. Inclined to work within the parameters of power relations as they find them, leaders of this ilk do little to challenge the securitization of socio-economic conditions by private actors. On the contrary, the discourse of extreme conditions is embraced as a way to connect with and influence mass publics in substitution for the ties of shared opinion. For executive authority in this mould, emergency politics is a way to foster acceptance towards unpopular policies and extensions of power, absolving their sponsors of some responsibility while suggesting that, before and after the exceptional moment, policy-making proceeds on a stable footing.

To highlight these factors is to look beyond some of the particularities of one episode. There is, of course, a fine-grained account to be given of the peculiar dynamics of the Eurozone, the bargains struck at Maastricht, and the choices of particular agents. Some of this historiography is already emerging; much remains to be written as records emerge. My intention has been to step back from accounts of this kind to examine some wider tendencies. In a political context where many are positively inclined to embrace the politics of emergency, the dysfunctionalities of particular policy regimes are a secondary issue. Too much emphasis tends to be put on the specificities of the Eurozone in this context.[46] There are, moreover, good reasons to avoid relying on notions of functional imperative—for instance, that political responses are determined by the need to take further steps of integration to correct existing failures.[47] Functional demands are relative to the ends one values, and therefore are a weak basis for explanation on matters where there is reasonable disagreement.

[46] On factors endogenous to EU institutions in the making of Eurozone and migration challenges: Schimmelfennig ('European integration (theory)...'), pp. 976ff. On how EU scholarship on Europe's economic malaise overplays the specificities of Eurozone design at the expense of wider and longer-term factors: Tooze (*Crashed*), e.g. p. 323; Sandbu (*Europe's Orphan*), p. 24.

[47] E.g. Philipp Genschel and Markus Jachtenfuchs, 'From Market Integration to Core State Powers: The Eurozone Crisis, the Refugee Crisis and Integration Theory', *Journal of Common Market Studies* 56 (1) (2018), pp. 178–96; see also accounts associated with 'post-functionalism' that identify matters as how EU policy-makers cope with functional demands for further integration in an age of popular dissensus (Hooghe and Marks, 'Cleavage Theory').

The argument implies that what we have come to know as the 'Euro crisis' may not be wholly distinctive and that emergency rule may recur—perhaps even that it is not yet concluded. I shall postpone the temporal dimension to the next chapter, which will examine in more detail the problem of endings and relapses. But it is important to ask at this point how the same dynamics might play out in different domains. Is the transnational politics of emergency essentially a phenomenon of matters economic, or can the scope of the argument be extended?

In the years around 2015 it became common to speak of a separate urgent challenge to the EU order. The reference-point for what became known as the 'migration crisis' was a spike in the number of migrants entering the EU from Africa and the Middle East in spring 2015.[48] Like economic policy-making, migration management connects to some of the core purposes of European integration. Free movement has been treated as central to the EU's capacity to build prosperity, as one of the four freedoms of the market. Insofar as the EU's founders sought to link it to goals of peace and security, cross-border movement has also been valued as something that contributes to the deconstruction of national boundaries. A 'crisis' in migration, whatever its form, is thus one potentially touching on the core goals and achievements of European integration. Problems arising here can challenge the polity's *raison d'être.*

The governing measures adopted in 2015–16 on migration showed some important differences from those in the economic domain. The latter drew in the whole range of EU states and supranational executives, both inside and outside the Eurozone. Involvement in governing migration was more uneven. With some parts of the EU more affected than others, some governments argued the need for swift action while others, notably the landlocked, hoped to sit things out. The response was more localized, the density of measures rather lower.[49] Nor was there a close equivalent to the ECB, i.e. a powerful supranational agent within the EU framework central to the crisis management. But one sees striking similarities nonetheless, especially in those later phases that involved not just unilateral measures by individual states but efforts to coordinate a response. One sees a politics escaping norms, rationalized by appeal to exceptionalism, and steered by shape-shifting executive power.

[48] On the dubious term 'migration crisis': Gurminder K. Bhambra, 'The Current Crisis of Europe: Refugees, Colonialism, and the Limits of Cosmopolitanism', *European Law Journal* 23 (5) (2017), pp. 395–405. On migration numbers: UNHCR (2015), 'Over One Million Sea Arrivals Reach Europe in 2015', 30 Dec.: http://www.unhcr.org/afr/news/latest/2015/12/5683d0b56/million-sea-arrivals-reach-europe-2015.html

[49] Agustín José Menéndez, 'The Refugee Crisis: Between Human Tragedy and Symptom of the Structural Crisis of European Integration', *European Law Journal* 22 (4) (2016), pp. 388–416; Genschel and Jachtenfuchs ('From Market Integration…'); Marco Scipioni, 'Failing Forward in EU Migration Policy? EU Integration after the 2015 Asylum and Migration Crisis', *Journal of European Public Policy* 28 (9) (2018).

As increasing numbers of refugees entered the EU in spring 2015, seeking asylum not in the country of arrival but onward passage, existing EU rules for handling inward migration centred on the Dublin system became subject to high-profile breaches. Mediterranean states encouraged new arrivals to continue their journey rather than seek asylum. Although a breach of positive rules, arguably these moves were in keeping with norms of humanitarianism to which the EU subscribes. More problematic were the steps that followed, premised on the threat the new arrivals were said to pose. Controls were reintroduced at the borders of several states, signalling the partial suspension of the Schengen system of free movement and an exceptional regime of uncertain duration.

As with the Euro crisis, responses soon extended beyond the national level. A 'temporary emergency relocation system' was agreed between EU states and introduced in September 2015 to control the distribution of asylum seekers, while parallel efforts were made to reduce the numbers of those entering the EU.[50] A notable feature would become the ad hoc involvement of third-party institutions, neither of the state nor the supranational sphere. Like the IMF's incorporation into the Troika, here an unconventional role was played by NATO, especially from spring 2016. Its involvement was partly to facilitate EU negotiations with Turkey, one point of departure for migrants. It was also, like the Troika, explained as a way to relay information.[51] But what made NATO especially attractive was that it lay outside the constitutional frameworks of the EU and its member states. It could be used to escape their legal and political norms.

Specifically, NATO's involvement cast policy as *foreign* policy, and allowed contentious goals such as deterring migrants at sea, along with delicate decisions concerning the distinction between migrants and refugees, to be outsourced to a military organization.[52] Norms to do with the reception of migrants, and more generally to do with the conferral of powers, could be set aside. An unconventional and evolving arrangement—legally questionable, yet also premised on taking advantage of the limits of European law[53]—became central to the crisis response.

Rationalizing such measures were appeals to the exceptional. Interventions were advocated on grounds of unique circumstance. One saw the evocation of quasi-natural forces—'unprecedented' and 'unexpected migratory flows', 'unpredictable migratory pressure'[54]–demanding a far-reaching response. Such an

[50] Arne Niemann and Natascha Zaun, 'EU Refugee Policies and Politics in Times of Crisis: Theoretical and Empirical Perspectives', *Journal of Common Market Studies* 56 (1) (2018), pp. 3–22.

[51] Andreas Jacobs, 'In Troubled Waters: NATO's New Maritime Activity in the Aegean', NDC Research Report, Research Division, NATO Defense College 1/16 (March) (2016), p. 4.

[52] Yuri Zhukov, 'NATO's Mediterranean Mission: What the Alliance Is Doing in the Aegean Sea', *Foreign Affairs* 21 February 2016.

[53] Though Menendez ('The Refugee Crisis', p. 401) calls it an assault on 'the very ideal of the rule of law'.

[54] European Commission (2016), 'Proposal for a Council Implementing Decision Setting Out a Recommendation for Temporary Internal Border Control in Exceptional Circumstances Putting the Overall Functioning of the Schengen Area at Risk'. Brussels (COM(2016)275), pp. 3–4. The CJEU

interpretation depended on emphasizing the novelty and suddenness of increased migration, and delinking it from long-term factors, for which responsibility might be shared, such as environmental change, economic inequality, overseas interventions, or weak statehood.[55] It also depended on emphasizing, in addition to humanitarian concerns, the threat posed to security, peace, and free movement. The partial suspension of Schengen was used to highlight that foundational principles of European integration were in question.

This was also the hour of the executive, as a constellation composed of those at the pinnacle of a dispersed set of institutions. Decisions were co-produced by the representatives of leading national governments, the Council and Commission, and, in addition to NATO, EU agencies such as FRONTEX and EUROPOL. They worked with speed and discretion to improvise a response, offering mutual validation of each other's claims to be acting under necessity.[56] As with the Euro crisis, these manoeuvres were designed to speak to multiple constituencies—the national populations of EU member-states, but also governments outside the EU and indeed migrants themselves. NATO's involvement helped perform a politics of speed.[57] Amidst this acceleration and informalization of executive power, parliaments were largely side-lined.[58]

Finally, these moves amounted to a potentially long-lasting redistribution of power. Certainly, many measures were said to be temporary, or geared to 'completing' existing systems. But the legacy was the development of new policy regimes and institutions for migration control. It was a form of polity-extension. Notable were moves to reconstitute FRONTEX as the European Border and Coast Guard (EBCG), with extra resources and expanded tasks. Amongst the proposed new capacities, and reminiscent of the Troika, were newly intrusive mechanisms to compel member-states to maintain border security, with outside officials overseeing compliance. The Commission's proposals of December 2015 capitalized on the months-long difficulties of the Schengen zone to pursue a transnational border

subsequently confirmed the migration movements amounted to an 'emergency situation' that was 'sudden', 'unforeseeable', and 'unprecedented': Judgment ECLI:EU:C:2017:631, 6 September 2017, Slovakia v. Council (http://curia.europa.eu/juris/celex.jsf?celex=62015CJ0643&lang1=en&type=TXT&ancre). On wider reliance on emergency framing: Daria Davitti, 'Biopolitical Borders and the State of Exception in the European Migration "Crisis"', *European Journal of International Law* 29 (4) (2019).

[55] For an empirical examination of these patterns in national and EU discourse: Pietro Castelli Gattinara ('The "Refugee Crisis" in Italy as a Crisis of Legitimacy', *Contemporary Italian Politics* 9 (3) (2017)), who concludes (p. 328): 'The state of exception is rapidly becoming the new rule to deal with migration in an increasingly militarised, law-enforcing and authoritarian fashion.'

[56] See e.g. the Commission publicly confirming 'the necessity and the proportionality of the measures taken' by Austria and Germany: European Commission (2016), 'Proposal for a Council Implementing Decision...'), p. 4.

[57] On NATO as a fast actor, and the speed of the decision to involve it: Jacobs ('In Troubled Waters'), p. 7.

[58] Menendez ('The Refugee Crisis'), p. 402.

regime.[59] Though not all measures have been enacted, those that have been are significant and potentially enduring.[60]

Not only was this emergency politics in a similar mould to the Euro crisis, but it was driven by similar factors. A concern to achieve and maintain order at the transnational level, sharpened by a keen awareness of its potential for breakdown, was an underlying sensibility in the internationalization of the response. The challenge of preserving certain core policies and purposes associated with the EU, to do with free movement in particular and continental security more broadly, gave executive agents some of the licence needed to override concerns about the means adopted.

The EU's diffuse institutional structure meant there was no single agent able to respond in an authoritative fashion, leaving space to be filled by several on a spontaneous basis. The same institutional structure created added scope for discretion: faced with more than one framework of rules, and weak codification of their relations, executives could be selective in which they adhered to, exiting where preferred into ad hoc agreements. These possibilities were actualized in the face of material pressures, amplified and securitized by those in the wider society. Proposals to strengthen the EU's border regime gained traction after widely mediatized scenes of chaos that seemed to put the Schengen zone's future in question. Images emphasizing the scale of migration, and commentaries on the 'end-of-Schengen', communicated the urgency of the moment and the sense it was unprecedented. Far-Right groups presented the movement of people as a crisis for receiving populations, while NGOs raised humanitarian concerns that, though very different in moral focus, could be appropriated by others and entwined with security discourses. By co-producing the sense of emergency, these agents helped shape the response.[61]

Finally, what of the ethos of executive power—was this another exercise in 'governance'? The moralized discourse of the German government when waiving the Dublin rules might suggest a less calculating outlook. But at base there remained something closer, visible in the allocation of decisions to unelected bodies and the use of managerial tools. As migration routes became formalized as 'corridors', they became objects of transnational governance, spaces in which

[59] European Commission (2015), 'Communication from the Commission to the European Parliament and the Council: A European Border and Coast Guard and effective management of Europe's external borders', Strasbourg (COM (2015) 673). On Commission representatives spotting an opportunity in a context of emergency: Arne Niemann and Johanna Speyer (2018), 'A Neofunctionalist Perspective on the "European Refugee Crisis": The Case of the European Border and Coast Guard', *Journal of Common Market Studies* 56 (1), esp. pp. 32–3; cf. Giuseppe Campesi, 'Crisis, Migration and the Consolidation of the EU Border Control Regime', *International Journal of Migration and Border Studies* 4 (3) (2018), pp. 196–221.

[60] Niemann and Speyer ('Neofunctionalist Perspective'), p. 27; cf. Marco Scipioni (2018), '*De Novo* Bodies and EU Integration: What is the Story behind EU Agencies' Expansion?', *Journal of Common Market Studies* 56 (4) (2018), pp. 773–4, pp. 776.ff.

[61] Davitti, 'Biopolitical Borders', p. 1186.

local rules might be suspended and intensive new sites of administration introduced. Potentially volatile migration movements that challenged sovereign control were thereby channelled and contained, and in a form that allowed networks of local, national and supranational authorities to be involved.[62] Also in the language of the crisis management—the appeal to 'risk analysis', 'vulnerability assessments', and 'migration management support teams'[63]—and in the focus on quantitative measures to define problems and solutions,[64] one saw traces of the governance outlook. Here were the fingerprints of those 'professionals in the management of unease' whose networks are observed increasingly to shape EU migration policy.[65] The evocation of threats, risks and vulnerabilities reflected the self-understanding of these officials as specialists in security, as those who could offer protection.

Finally, one should not overlook that one of the effects of focusing public attention on questions of migration was to shift it away from malaise in the economic sphere. By addressing themselves to a new set of challenges, executives were at the same time addressing the weakness of their problem-solving capacity elsewhere. The events of the 2010s are often narrated as the unfortunate coincidence of parallel crises, but the interconnections deserve emphasis. As the events of the Euro crisis reaffirmed the limits of output legitimacy, so they sharpened the appeal to policy-makers of a politics rooted in ideas of threat, necessity, and last resort, for which the field of migration offered striking material. After a period of questionable results, here was an opportunity for executive agents to reassert their credentials as problem-solvers.

What emerges from these events spanning economics and migration is a common set of underlying truths. Upholding the EU's policy commitments—the freedoms of the market, the movement of persons—has always been a challenging proposition, more so in view of societal changes that question the capacity of institutions to call the shots. Policy-makers sought to adapt by embracing a problem-solving ethos that licenses discretion, unmoored from political and legal restraints. The results have been decidedly mixed. Faced with the ongoing challenge of getting things done, and aware of their diminishing ties with a wider public, executives in the EU came to emphasize the determining pressures of the moment, drawing on themes of urgency and necessity. They embraced a mode of rule avowedly reactive in nature, centred on responding

[62] On corridors and the establishment of governing 'hotspots': Davitti, 'Biopolitical Borders'; Reece Jones, Corey Johnson, Wendy Brown, Gabriel Popescu, Polly Pallister-Wilkins, Alison Mountz, Emily Gilbert, 'Interventions on the State of Sovereignty at the Border' *Political Geography* 59 (2017).

[63] Commission press release, 'A European Border and Coast Guard to Protect Europe's External Borders', p. 1.

[64] See e.g. Juncker's state of the union address, 9 September 2015: http://europa.eu/rapid/press-release_SPEECH-15-5614_en.htm

[65] Didier Bigo, 'Security and Immigration: Toward a Critique of the Governmentality of Unease', *Alternative* 27 (Special Issue) (2002), pp. 63–92.

quickly to material pressures. In the transnational sphere they found macro-threats to engage, and monitoring mechanisms and expert networks to reinforce them. Policy got made, in some fields more decisively than others, but generally as something required rather than desired. Both in its self-understanding and its wider context, this was a politics of last resort.

Conclusion

One way to study the EU is under a microscope, as a series of concrete negotiations and decisions, of policy successes and failures. Much that has been written on the upheavals of recent years follows this route, examining the origins of the Euro crisis in decisions about Eurozone design made in the 1990s. Here I have taken a different path, seeking to account for the practices of emergency rule in factors not limited to any one policy regime or set of events.

Whereas the previous chapter argued that the transnational sphere has long had vulnerabilities to this mode of rule, here I have indicated why executives more recently have become inclined to take advantage. The transnational politics of emergency emerges from a series of challenges to political authority. The capacity to define socio-economic conditions as needing a dramatic response is increasingly shared by a wide range of agents beyond executives themselves, ones structurally on the look-out for extreme conditions. The dependence of political authorities on their data and evaluations gives ample scope for the application of pressure. Changes in the make-up and outlook of executive power encourage such pressures to be accepted in the name of problem-solving rather than ignored or resisted. This ethos of governance is especially developed in supranational institutions, but finds expression in state circles too. Its origins are broader than the influence of any one ideology.

The politics of the Euro crisis remains the centre of our discussion, but the response to migration in this period also bore features of transnational emergency rule. Here too one saw steps towards further European integration taken as emergency measures. Polity-building and emergency politics went hand in hand. Moreover, these measures in the migration domain were driven by factors resembling those in the economic. Indeed, separating the two into discrete episodes is arguably something to avoid. Part of the logic of emergency rule is to cast historical events as self-contained, each set of measures a unique response to unique circumstances. A critical reading demands one recognize the factors that link them. The substantive nature and scale of the extreme conditions may vary, but the responses arising are expressions of the same governing template.

4
Prospects of a New Normal

The idea of a state of emergency implies the possibility of isolating a period of time for a distinctive form of rule. The suggestion is that the usual modes of government—the procedures that compose them, the norms they instantiate—are inadequate for dealing with an extreme set of circumstances. In all cases the logic of the claim is that there is a beginning and an end: something called normal rule is merely interrupted. While the intervening period clearly should have *some* consequences for the period of normality expected to follow—commonly it is said to make the latter possible—it is not expected to merge with it or set precedents for it. What happens in the exception stays in the exception.

In the first chapter I described this idea of normality-to-come as the 'posterior normal'. It is not about reverting to the status quo ante. Though it may involve discontinuing certain measures, equally it may involve retaining those considered necessary to prevent a return of crisis conditions. Rather than the restoration of an anterior normal, the promise of normality is a more general one: that the polity's most basic norms and procedures will be binding in future.[1] What, we may then ask, must be true for this to be a realistic proposition, something more than a merely ideological assertion designed to minimize discontent?

This is a matter of the procedural authority that the polity can carry. Later in the book we shall look at forms of authority grounded in technocracy, where expertise is the reference standard—a natural one when executives are engaged in unconventional practices that make recourse to procedural authority difficult. But rule detached from the authority of established norms always risks becoming indistinguishable from arbitrary and authoritarian rule—rule that, while it may sometimes deliver acceptable results, does not do so on any predictable and responsive basis.[2] Hence (re-)establishing procedural authority is likely to be a goal for executives and others alike.

At stake would seem to be a successful act of partitioning and closure, such that the extraordinary measures of emergency rule do not contaminate the institutional

[1] Bindingness is a core element of the rule of law (cf. Brian Z. Tamanaha, *On the Rule of Law: History, Politics, Theory* (Cambridge: Cambridge University Press, 2004)), but as will become clear my focus is broader.

[2] These broad observations are inspired by the account of republican freedom provided in Quentin Skinner, *Liberty before Liberalism* (Cambridge: Cambridge University Press, 1998). The felt obligation to comply with a norm, though it may have a range of moral sources, is typically nurtured by belief in its capacity to constrain.

Politics of Last Resort: Governing by Emergency in the European Union. Jonathan White, Oxford University Press (2020).
DOI: 10.1093/oso/9780198791720.001.0001

regime that ensues. The period of emergency needs to be delimited, so the irregularities it entails do not fatally undermine the polity's norms and procedures more generally. To achieve this division in a context such as the EU, where emergency rule lacks legal codification, the question is in large part a political one. Decision-makers must make a convincing case for why they will not resort to extraordinary measures again.

Specifically, I suggest, one of three claims would need to be credibly advanced. The first would be that crisis conditions will not recur, perhaps because their causes have been dealt with, or because better policy-making will contain the problems arising. There will therefore be no *need* to resort to extraordinary measures in future. The second kind of claim would be that, should crisis recur, new procedures are in place that will minimize executive discretion. There will therefore be no *scope* to resort to extraordinary measures. A third claim would be that the identity of the decision-makers has changed, such that those with a proclivity for exceptionalism have left the political scene. There will therefore be no *inclination* to resort to extraordinary measures.

The chapter examines what would need to be true for such claims to be plausibly advanced. The analysis suggests that they are difficult to make credibly in the contemporary EU. An emergency regime is difficult to terminate. As I go on to argue, persistent weakness in the polity's norms must be reckoned with. Those suspended, violated, or evaded, as well as others not directly challenged, are put in question by emergency rule and invite scepticism concerning their capacity to bind. Procedural forms of authority emerge from emergency rule in poor shape.

The final section discusses some implications. To the extent that drawing a line under emergency rule is difficult, ideas of a normality-to-come will be misleading. They may have some purchase on the public imagination—arguably the promise of an ensuing period of calm is part of the symbolic power of emergency politics—but any normalization that occurs is likely to be superficial. Unable to rely principally on procedural forms of authority, executive powers are likely to combine recurrent appeals to the logic of emergency with appeal to alternative sources of authority and to governing by coercion. Machiavelli's warning was well taken: 'in a republic, it is not good for anything to happen which requires governing by extraordinary measures. Although extraordinary measures may be beneficial at a certain moment, the example nevertheless causes harm, because if one establishes the habit of breaking the laws for good reasons, later on, under the same pretext, one can break them for bad reasons.'[3]

[3] Machiavelli (*Discourses*), I.34, p. 95.

The Problem of Reassurance

All major ruptures and interruptions in political time pose a challenge for the period that follows. What confidence can there be that the extreme discretion previously exercised will not be a persistent feature of the resulting regime? Political revolutions are perhaps the most dramatic form in which the question is posed, but it arises on a spectrum of transformations that include regime changes and negotiated secessions.

We can think of this as the problem of *constitutional reassurance*. Often it is approached as a matter of the restoration—or consolidation—of the rule of law. The large literature on law in 'transition' societies can be read as an engagement with the problem of reassurance.[4] Here a special role tends to be accorded to courts in identifying and nurturing the coherence of the legal order, upholding it against transgressions. Restoring legal continuity with a previously existing constitution, or protecting the integrity of the new one, are some of the central tasks. But building the credibility of a legal order is just one dimension of constitutional reassurance. Equally at stake is confidence in those conventions of practice that go beyond what is legally codified yet on which the functioning of a polity depends. Reassurance entails a more general act of persuasion that whatever period of unconventional rule was previously manifest is now past.

Constitutional reassurance is not necessarily something everyone may wish for. Consider those who suspect that the legacy of a certain revolution is an order that is fundamentally unjust. The consolidation of this order and the general acceptance of its procedures may be exactly what they fear. Terms such as normalization may be used to describe the unwelcome prospect that past transformations are absorbed into the new order while the possibilities for future transformations are diminished. But for those who have a stake in the emerging order and wish to see its consolidation, a central goal is likely to be to build public confidence in the post-crisis order, such that it is viewed as something more robust than what existed hitherto.

Emergency rule is one of the phenomena that pose the problem of reassurance with some force. For the sake of dangers often invisible or unsusceptible to everyday experience, it disrupts the formal divisions of power supposed to characterize the polity and the norms these plausibly serve. Since executive discretion is a provocative way to govern, the accompanying suggestion is that these are merely temporary disruptions. The logic of emergency is that a return to more bounded forms of rule is imminent. It relies on the idea that constitutional reassurance is possible. But how then is the implied act of separation between the exceptional and the normal to be made?

[4] See e.g. Ruti Teitel, 'Transitional Jurisprudence: The Role of Law in Political Transformation', *Yale Law Journal* 106 (1997).

There may be contexts where such questions can be evaded. Where the application of extraordinary methods can be concealed, the wider legacy of emergency rule may be limited. Where breaches of norms can be done covertly, and innovations presented as upholding the existing order, larger questions of authority may be avoided.[5] But exceptionalism cannot always be hidden—in particular where its effectiveness is thought to depend on its visibility. Sometimes true in conventional military states of emergency, this was certainly the case with the politics of the Euro crisis. The logic of reassuring the markets seemed to require not only that extraordinary measures be taken but demonstrably so. Executives tended to be explicit about the degree to which they were exceeding existing norms: there was little pretence to 'business as usual'.[6] This is important because the visibility of norm departures is arguably more significant for their subsequent credibility than the fact of departure as such.[7] Generally speaking, a normative order can withstand plenty of evasion, but will do so more easily where these challenges are not widely recognized. Where exceptionalism has been open for all to see, it becomes harder later on for executive powers simply to deny its occurrence. Rather, the bindingness of norms must be re-established. The problem of reassurance becomes hard to avoid.

In a well-developed constitutional order, it may be that the delimitation of emergency rule can be performed by institutional means. Establishing two identifiably distinct governing regimes, one associated with emergency rule and one with normal conditions, is something many states have sought to do.[8] This provides a solution of sorts to the problem of reassurance. Formally discontinuing an emergency regime is one way to signal the restoration of constitutional norms. Even this, however, is a quite limited solution to the problem. It does not exclude that a return to emergency rule is imminent. Moreover, empirical studies suggest the boundary between normal and emergency regimes is often blurred. Whereas the ideal-typical emergency regimes of ancient Rome were clearly circumscribed institutionally, emergency power being concentrated in a designated individual, this formal clarity is rare in the modern world.

In any case, in the EU context, this institutional solution is presently unavailable. The kind of emergency rule we have been examining had a distinctly informal quality. It did not have its own defined set of practices, and certainly

[5] On undeclared states of emergency: Gross and Ní Aoláin (*Law in Times of Crisis*), pp. 305, 318–19.

[6] This instance of emergency rule resembles the so-called 'accommodation' model—i.e. selective suspensions, violations and improvisations around the margins of an ostensibly stable legal framework—rather than 'business-as-usual' legality or the wholesale interruption of legality (cf. Gross and Ní Aoláin (*Law in Times of Crisis*), chaps 1–3).

[7] On the loose relationship between practices and norms: Brennan et al. (*Explaining Norms*), pp. 20ff.

[8] Cf. Gross and Ní Aoláin (*Law in Times of Crisis*), pp. 174ff. On the 'emergency state': Aaron Klieman, 'Emergency Politics: the Growth of Crisis Government', *Conflict Studies* 70, April (1976).

no constitutional codification thereof.[9] Nor, as we have seen, were the norms and procedures of the preceding order well codified. Consequently, there is no straightforward institutional means, however imperfect, by which to discontinue emergency rule and offer constitutional reassurance.

The act of separation between exceptionality and normality must therefore take the form of a claim. Decision-makers must credibly advance some type of account of how emergency rule has been superseded, to be replaced by a new period of normality in which procedures and norms are binding. The following analysis highlights some of the possibilities available for making a claim of this kind, based on general features of the structure of emergency rule, and with particular reference to the challenges posed in the transnational sphere.

Again, I do not want to suggest that all relevant parties will necessarily share the goal of wanting to stabilize the post-emergency institutional order. As noted, those fundamentally opposed to the status quo will see this as a conservative goal. Equally, the political realists amongst established powers, though they may not share this fundamental dissatisfaction with the status quo, may feel that emergency rule is a perfectly acceptable response when circumstances of necessity demand it, and that there is therefore no reason to want to exclude its possibility. Emergency rule has its place, they will say, and those seeking constitutional reassurance are naive.

These are some of the crucial normative questions surrounding the transnational politics of emergency, and we shall discuss them further in Chapter 7. But provisionally we may assume that a desire to rule out the imminent prospect of further outbreaks of emergency discretion is a goal likely to be shared by many.

From Discretion to Procedure?: On Concluding an Emergency Regime

I suggest there are three main ways in which the claim to have left emergency rule behind might be made. Let us consider them in turn. The first (1) rests on the notion that the relevant *extreme conditions have been overcome*. It is claimed that the conditions of crisis that made exceptional measures necessary have passed and are unlikely to return for the foreseeable future. The threats have subsided. There will hence be no renewed need for suspensions and improvisations under the sign of emergency.

[9] The special EU/IMF lending programmes associated with the 'bail-outs' mark an institutional separation of sorts, and their conclusion is sometimes heralded as the end of emergency measures. These do not amount to well-established procedures however, and what follows them remains uncertain.

Such a claim might rest on the idea that circumstances have changed, such that the causes of extreme conditions have receded. Much depends on the domain of experience in question. What tend to be called 'natural' disasters such as earthquakes and floods would seem to consistently resist this view, insofar as they recur, and recur on an unpredictable basis. But in fields where the forces at work ebb and flow, or where they can plausibly be tamed by human action, such an interpretation may in principle be possible. A different version of the claim might rest on the notion that today's policy-making is qualitatively better than yesterday's at solving the relevant problems as they arise, so that even if the same challenges are met, decision-makers will be better able to handle them. A superior order has been instituted, better attuned to the challenges faced. The political order's norms should be regarded as credible because there will be no need to interfere with them in future.

Clearly one of the challenges for this line of argument is how far one can ever credibly allay fears of a return of crisis conditions. Even policy-making that is in some sense 'optimal' is such only for a given set of circumstances and relative to certain values. To the extent that these background conditions inevitably change in unforeseeable ways, one can expect situations of maladjustment and stress to recur. This holds especially true if efforts to deal with previous crises sow the seeds of future instability. It is not unusual for emergency measures to draw on the same ideas that the problems they face put in question. The remedy extends the disease. Where this is true, assertions that extreme conditions have been left behind must ring hollow.

Such a claim also risks ignoring the ambiguity of what constitutes a crisis. Executive discretion in the name of urgent threat is never born solely of necessity: it depends on actors' willingness to identify certain conditions as those of a crisis, and the willingness of others to accept this diagnosis. Historians tell us that, even in apparently clear-cut situations of upheaval such as a civil war, the definition of a period of emergency is a delicate and contentious matter.[10] Given those who have engaged in emergency politics may have derived advantage from doing so, there will be a persistent temptation to repeat it, irrespective of objective circumstances and notwithstanding the interim benefits of a more settled form of government. A claim based on system capacity and a change in environmental conditions misses this important point. Moreover, in the economic domain as we have seen, definitions of crisis are also heavily shaped by powerful private actors, including investors, speculators, market analysts, and ratings agencies. Given their role in co-producing and announcing the crisis, one will need to be cautious regarding

[10] On the temporal delimitation of emergency in the American Civil War: Sabine Mischner (online early), 'The Temporalities of Exception. The Long Shadow of the American Civil War', *Zeitschrift für Politikwissenschaft* 3 (2018); cf. Stephen Holmes, 'In Case of Emergency: Misunderstanding Tradeoffs in the War on Terror', *California Law Review* 97 (2) (2009), pp. 301–55.

the capacity of executives to forecast the return of peaceful conditions. Still, this is clearly one path that can be pursued towards constitutional reassurance.

A second type of claim (2) hinges less on the external conditions and more on the institutional response. Reassurance may be sought with the proposition that there are now *new constraints on executive power*. Such a claim would hold that, although the prospect of crisis persists, when it comes it will not be handled with resort to far-reaching executive discretion because a new framework has been instituted that renders this unnecessary or unfeasible. The post-crisis order, it may be said, is better insulated from executive interference. On the grounds that there will be no unconventional manoeuvres next time, procedural authority is affirmed.

The credibility of such a claim would seem to hang first on whether serious efforts are made to regularize the decision-making structure. The accumulation of emergency measures tends to leave a fragmented institutional terrain. Either because they have been designed to allow decision-makers flexibility of action, or because, introduced hastily, they clash with and duplicate one another, emergency provisions are unlikely themselves to be a reliable constraint on the future exercise of discretion. Legal principles such as the primacy-of-EU-law will be left disrupted by the resulting pluralization of legal forms, and the level of opacity these arrangements produce allows further freedom from scrutiny. A convincing claim to have moved beyond executive discretion thus depends on moves to repeal emergency provisions or integrate them into a new, more regularized structure.

More generally, a convincing claim to have limited the capacity for executive discretion is likely to depend on the visible empowerment of other institutions. It might hinge on legislatures being given greater powers of decision-making, such that executives cannot discount their positions. Equally, it might involve granting judiciaries greater powers, either to make executive discretion more difficult to enact, or to discourage it with the threat of sanction.[11] In such ways, procedural constraints on decision-making might ward off the prospect of renewed exceptionalism in future.

Note that this is not principally a question of the retrospective legal recognition of emergency measures. Certainly this may assuage some of the more flagrant tensions in the post-emergency order and bring them into the legal framework. But on its own it does little to offer constitutional reassurance, as it does little to ward off the possibility of similar such moves in future. (Indeed, the sceptic would say that it invites them.) The risk is that it achieves an 'authoritarian legality' that makes the persistent problems of the association harder to identify.[12]

[11] On some of the possibilities here: Oeter et al. (*International Law*).

[12] R. Diab, *The Harbinger Theory: How the Post-9/11 Emergency Became Permanent and the Case for Reform* (Oxford: Oxford University Press, 2015), pp. 86–93.

Recall from Chapter 2 one of the structural problems of the transnational sphere that any claim to constrain executive power must deal with. Executives have the option to go outside one particular framework (like the EU's) to make arrangements based on international law. A kind of 'forum shopping' is available. Should they wish again to resort to ad hoc measures of decision-making—and it may suit powerful member-states well to do so[13]—what will prevent them? One can assume that significant changes in the EU's structure would be needed for its norms to be less susceptible to suspension or marginalization than before.

There is a third path to constitutional reassurance one should consider. This is one built on the claim (3) that *the identity of the decision-makers has changed*. It might take as its starting-point the idea that it will never be feasible to design institutions such that exceptionalism is impossible: rather, everything depends on the inclinations of the individuals in charge. Some will seek to maximize their discretion; others will seek to handle a crisis by conventional means, and where revisions are necessary will submit these to a public process of deliberation rather than force them through as an urgent necessity. Following this line of argument, the reason to approach a polity's norms with renewed confidence is that those who (ab)used discretion in the past will not be in a position to do so again. To the extent that they no longer hold office, a reversion to their methods is not to be expected.

Those advancing this type of claim will have to downplay the possibility that a quite different set of executive agents will still be inclined to behave the same way. If there are strong pressures emerging from the socio-economic sphere that impel executives towards discretionary action, and little opportunity for them to be held accountable for their decisions, there may be little reason to attach significance to a change in their individual composition. Executive agents will arguably do as they do, irrespective of the persons involved. Still, it might be said that there is a learning process at work, and if those coming to power see that their predecessors have suffered electorally for their involvement in emergency measures then they have reason to act differently. A new cast will perform their roles differently.

This type of claim is more easily advanced in the context of a democratic state. The challenge in alternative settings is the lack of institutional mechanisms by which a change of personnel can be heralded. Elections are typically the means by which the ritual of catharsis is performed in political life. Yet traditionally there has been no opportunity to expel in one go the principal executive actors of the EU. Many offices are appointed on a long-term basis, often without an electoral component. More generally, the co-production of emergency rule by a wide array of agents distributed across many sites of institutional power raises the prospect that the turnover of personnel is slow and piecemeal.

[13] Henning (*Tangled Governance*).

Emergency Rule Concluded?

How then does the EU fare after its involvement with emergency politics in the 2010s? How credibly can such claims be advanced? The first—that extreme conditions have been overcome—is one that will be familiar to those who followed the political economy of Europe in this period. Claims to this effect were made by many of those wielding executive discretion. They invoked the goal of exiting an extraordinary period and returning to one more ordered, centred for example on remedying the design flaws of the Eurozone.[14] It was sometimes explicitly said by leading officials that the threats to which extraordinary measures responded had subsided.[15]

But there are reasons for cautious appraisal. Some follow from points already made. Measures instituted to handle the economic upheaval of the period arguably served to embed some of the drivers of economic volatility. Note for instance how the austerity policies and budget controls adopted themselves increase the chance of future crises in the Eurozone and beyond.[16] Such policies arguably make it more likely rather than less that executive agents in future will cite extreme economic conditions as reasons to escape the hold of political and legal norms. The noted ambiguity in what constitutes circumstances sufficiently destabilizing adds further to the scope for it.

Some of the important contemporary trends in governance practice in the EU and beyond presuppose that crises *will* recur. Whatever leading officials may say publicly to cultivate the sense of a return to normality, scholars observe the rise of a logic of 'preparedness' in policy planning. Unlike approaches based on the rationality of risk, which aim to quantify and reduce the likelihood of extreme conditions, this approach sees them as largely unavoidable and aims instead to be well prepared when they arrive.[17] Such an outlook is partly the legacy of past episodes of crisis management. An effect of emergency rule is arguably to cultivate an attitude of anticipation amongst decision-makers concerning the *next* set of threats to be engaged. As an observer of US emergency government observes, one

[14] See Commission President Barroso's valedictory speech to the European Parliament, 21 October 2014, Strasbourg: 'we are stronger, because we have a more integrated system of governance, because we have legislation to tackle abuses in the financial markets, because we have a much clearer system of supervision and regulation'.

[15] Advancing this claim: European Council President Van Rompuy in his March 2014 report 'The European Council in 2013' (http://www.consilium.europa.eu/uedocs/cms_data/docs/pressdata/en/ec/141640.pdf, Luxembourg: Publications Office of the EU), where he writes of 2013: 'market tensions abated during the year, and we could safely say that the existential threats from the financial crisis were now firmly behind us'.

[16] Blyth (*Austerity*); Tooze (*Crashed*); Wolfgang Streeck, *How Will Capitalism End* (London: Verso, 2016).

[17] Ben Anderson, 'Preemption, Precaution, Preparedness: Anticipatory Action and Future Geographies', *Progress in Human Geography* 34 (6) (2010), pp. 777–98; Andrew Lakoff, 'Preparing for the Next Emergency', *Public Culture* 19 (2) (2007), pp. 247–71.

sees a shift from 'temporary crisis governance' to 'preparatory emergency management' as experience with the practice deepens.[18]

EU policy-making is now so shaped by the anticipation of future crises that some political scientists speak of its 'crisification'.[19] Finding the next urgent event is said to be a principal concern of officials and one expressed in a variety of institutional innovations. 'Horizon-scanning' systems, 'early-warning' systems, 'rapid response' mechanisms, and 'crisis rooms' have been developed in the Commission and beyond, and now deal with an expanding set of policy concerns. Not only do such mechanisms embed the exceptionalist outlook in the heart of decision-making, but they combine with the prioritization of speed of decision-making over competing concerns to do with procedure, deliberation, and participation.[20] Such practices suggest there is little institutional commitment to the notion that extreme conditions have been overcome and that crisis government is something of the past.

Indeed, one may suppose that certain crisis measures of the 2010s have been designed to allow emergency claims credibly to be advanced in future. What one might call the *heresthetics of emergency* involves structuring institutional arrangements such that future appeals to emergency will have plausibility.[21] Identifying such intentions is difficult but the possibility deserves mention. The continued quantification of policy-makers' exposure to market pressure, and very public references in this regard to such devices as 'the spread', notably allow the prospect of a loss of market confidence to be used to steer decisions at high-stakes moments. When Italian President Mattarella pressed the newly-formed Italian government in May 2018 to reaffirm its commitment to Eurozone membership by appointing a different Minister of Finance, he was able to cast the alternatives as unthinkable by invoking the 'alarm among investors and savers', underlining in particular 'the rising spread, day after day' and 'the losses on the stock exchange, day after day'.[22] The rhetoric had power because of the structure of the situation and the instruments available for its evaluation. Again, such episodes suggest the politics of emergency remains an enduring possibility.

What then of the second type of claim, that there are now *new constraints on executive power*? One need only look at the governing arrangements of today's EU to see such a process is doubtful. Certainly one sees moves to regularize some of the institutional innovations of the Euro crisis.[23] Possibly one even sees the

[18] Cf. Curley ('Models of Emergency') on the twentieth-century US experience.

[19] Mark Rhinard, 'The Crisification of the EU', *Journal of Common Market Studies* 57 (3) (2019), pp. 616-633.

[20] Rhinard ('Crisification').

[21] The term 'heresthetics' is from William Riker, *The Art of Political Manipulation* (New Haven: Yale University Press, 1986).

[22] For the president's statement: http://www.quirinale.it/elementi/1417

[23] See in particular moves to bring the ESM into the EU framework, and to establish a European Monetary Fund.

formation of permanent institutional structures that would render ad hoc formations such as the Troika technically superfluous in future. This follows the template of 'preparatory emergency management' as described.

But there are several reasons for caution. Many of the structures arising seem actually to *embed* opportunities for the future exercise of executive discretion. Ostensibly firm rules are defined, but with provisions for the Commission and Council to overlook them should 'exceptional economic circumstances' warrant it.[24] What constitutes exceptional circumstances is itself often dependent on their own empirical readings. Those who might identify the exceptional situation, and those who might act upon it, are generally therefore one and the same. Rather than the revocation of emergency rule, this looks like its enduring possibility.

Moreover, insofar as new restraints on executive power have been conceived in the EU, these have tended to involve a diversity of executive actors constraining each other. The terms of the European Semester entail oversight by the Commission and Council, influenced in turn by other executive agencies such as the ECB and IMF. The role of the CJEU remains uncertain, especially given the latitude for interpretation accorded to the Commission, and the Court's traditional weakness in the macroeconomic domain. While the discretion of specific governments may be constrained, executive discretion in general remains the system's defining feature.[25]

Scholars of EU law observe the marginalization of judicial power, at least as something that might challenge rather than corroborate executive power. Though the actions of governments and supranational institutions have not gone unopposed, judicial processes are slow compared to the actions they would challenge. Too delayed to fundamentally alter the course of events, judicial decisions have often served simply to give them the stamp of law. As with the constraints on legislative power, the effects on judicial power are potentially long-term. The new policy regimes and rules set up by executives in the crisis period typically rest on non-judiciable concepts that render judicial oversight all but impossible. If one component of constitutional reassurance has to do with

[24] Cf. EU Reg. 1174/2011 'on enforcement measures to correct excessive macroeconomic imbalances in the euro area', esp. preamble 15 and Article 4 (6), where the 'exceptional economic circumstances' formula is used. On the expansion of executive discretion, see Daniel Seikel, 'Flexible Austerity and Supranational Autonomy: The Reformed Excessive Deficit Procedure and the Asymmetry between Liberalization and Social Regulation in the EU', *Journal of Common Market Studies* 54 (6) (2016), p. 1408): 'the theoretical discretion of the Commission is exceptionally high. The reversion of the decision mode provides the Commission with an almost autonomous executive decision-making power for the surveillance and correction of national budgets.'

[25] There is some evidence of the strengthening of national parliaments in the 2010s, particularly—though not only—the *Bundestag* (Deirdre Curtin, 'Challenging Executive Dominance in European Democracy', *Modern Law Review* 77 (1) (2014), 25ff.), but the process is an uneven one, and the capacity of parliamentarians to exercise their new powers is questionable (cf. John Erik Fossum and Agustín José Menéndez, *The European Union in Crises or the European Union as Crises?*, ARENA Report No 2/14 (Oslo: ARENA, 2014), p. 16).

fostering the rule of law, the verdict of contemporary observers tends to be sceptical.[26] Moreover, exiting the EU framework should its demands be undesired remains very much a possibility under the existing model.

Finally, what of the third claim—that *the identity of the decision-makers has changed*? Clearly the truth of such a claim depends partly on when it is made. But the basic structural factors that obstruct it seem very much ongoing. There is no opportunity to expel in one go the principal executive figures of EU politics. The offices of the ECB are appointed on a long-term basis (eight-year terms for the presidency), and without an electoral component. The Commission's connection to electoral politics is weak, and even when alternative presidential candidates are presented in EP elections this offers limited capacity to a deliver a verdict on its actions in the round. Nor is the Council directly elected: its composition changes on what is—from a European perspective—an arbitrary basis, since it follows the varied rhythms and concerns of national politics. There is no definitive moment at which its make-up can be overhauled.

The Parliament certainly does have a basis in European elections, and on some readings has become a more powerful and assertive actor in recent years.[27] Still, it remains on balance one of the weaker EU institutions, certainly relative to the Council. To vote out its incumbents is hardly to vote out those responsible for the exercise of emergency rule. It sseems difficult then to draw a line separating the emergency regime from something that might succeed it. For an indeterminate period of time, the same people are responsible for both. The nomination (July 2019) of Christine Lagarde to become head of the ECB after eight years leading the IMF is one striking example of continuity of personnel amongst the leading transnational executives.

In sum, given broadly the existing configuration of the EU, credibly drawing a line under past periods of extraordinary rule is very difficult. The join between the emergency regime and its putative successor is seamless. A considerable portion of the former outlives the crisis, yet with little to suggest it will not itself be subject to discretionary intervention in future.

It is worth underlining that, under changed institutional conditions, stronger claims about the conclusion of emergency rule might be possible. An EU displaying stronger parliamentary powers would permit more convincing assertions concerning the restraint of executive discretion and the changeover of governing officials. An EU that fully revoked its commitment to austerity policies might

[26] Mark Dawson, 'The Legal and Political Accountability Structure of "Post-Crisis" EU Economic Governance', *Journal of Common Market Studies* 53 (5) (2015), pp. 976–93; Menendez ('Crisis of Law'); Joerges ('What Is Left').

[27] For analysis: Berthold Rittberger, 'Integration without Representation? The European Parliament and the Reform of Economic Governance in the EU', *Journal of Common Market Studies* 52 (6) (2014), pp. 1174–83; Desmond Dinan, 'Governance and Institutions: The Unrelenting Rise of the European Parliament', *Journal of Common Market Studies* 52 (2014), Annual Review, pp. 109–24.

permit more credible claims concerning the future avoidance of crisis conditions. Efforts to reorder the EU in such ways would invite a revised assessment. Constitutional reassurance is not a utopian idea, but in the circumstances analysed, each of the three claims seems problematic.

Enduring Problems of Procedural Authority

Given the noted difficulty politicians and officials face in convincingly consigning emergency politics to the past, it seems clear that procedural authority must remain in question. If such authority depends on people having good reason to see procedures, rules and norms as generally binding, the open-endedness of emergency rule poses a fundamental challenge, as it makes it hard to dismiss the prospect of their being manipulated once more. Hard to reasonably sustain is the belief that they are stable and general commitments reliably adhered to. Such an appraisal invites itself particularly for those interested parties who might be called upon to respect those constraints, including governments, who, taken individually, are objects as much as authors of these frameworks. Ordinary citizens too have little reason in these circumstances to approach the norms of their political order with confidence.

Most obviously in question are the specific rules, procedures and norms breached or suspended during the period of emergency politics. The suspicion must be that if this fate can befall them once it can do so again. Norms to do with state sovereignty, the equality of states and democratic control, and the wealth of institutional procedures and routines that may be said to serve them, look persistently vulnerable in this regard. If a repeat of their violation can hardly be excluded, their credibility remains tarnished. There can be little reasonable basis for a belief in their capacity to constrain.

If such a fate befell only those particular features suspended or transgressed, the significance would be local overall. An experimentalist rationale for the polity might still be possible: it could be defended as a form of spontaneous order, evolving by trial and error as much as conscious direction.[28] But it is never just those directly challenged that are affected. The binding character of ostensibly unrelated ones—i.e. those neither breached nor evaded—is also at stake. The larger constellation of norms is put in question.[29] This follows partly from the way political improvisation generally leads, as noted, to the accumulation of ad hoc measures which stand in a coherent relation neither to each other nor to the institutional structures that preceded them. Adopted swiftly for short-term goals

[28] Friedrich von Hayek, 'Rules and Order', vol. 1 of *Law, Legislation and Liberty* (London: Routledge & Kegan Paul, 1973).

[29] Cf. von Bogdandy and Ioannidis ('Systemic Deficiency').

and often with little deliberation, these measures may duplicate existing provisions, challenging the integrity of the wider framework.

It follows also from the simple observation that if executives are willing to intervene with discretion in one area they may be willing to do so in another. The limits to political discretion become difficult to establish with confidence. The point is not, of course, that the majority of norms will necessarily be subject to emergency-led suspensions, amendments and so on. Bureaucracies depend on a measure of continuity, in the EU no less than elsewhere. But though the exercise of such discretion may be limited in scale, the difficulty faced is how to know which norms are 'real' and which merely apparent. Clearly not all are equally likely to be transgressed: it is not a random matter, and an analysis of political and socio-economic power structures may provide some clues. But the course of emergency rule is sufficiently unpredictable that few elements of a polity's legal and political architecture can be regarded as secure. Not knowing which norms can be relied upon and which should be discounted, the observer must direct a measure of doubt towards all. They are interdependent as concerns their credibility.

Note also that even where extraordinary measures do not explicitly contravene existing EU commitments—e.g. because they are produced in heads-of-state meetings outside the EU framework—they may nonetheless put these latter in doubt, since they suggest their limited capacity to offer guidance in critical situations.[30] When push comes to shove, the rule of persons is seen to displace that of rules. Perhaps one should not be too surprised—it is arguably one of the tendencies latent in any political order—but it is especially problematic for one that has relied heavily on ideas of integration-through-law to legitimize itself.

One might suppose the aspects of the political order spared this subversion are those that emerge for the first time in the context of emergency rule. These, after all, are yet to develop a record of suspension. Yet they too must have question marks against them due to the conditions of their emergence. Measures taken under the auspices of emergency are inevitably associated with haste rather than reflection. They are likely to be tied closely to the details of the moment, in a way that reduces their generalizability and long-term validity.[31] Often they are intended to solve immediate problems rather than to stake out context-independent principles: for this reason they may become quickly obsolete, generating the need for further extraordinary measures, which in turn require replacing too.[32] This piling up of problem-solving measures also introduces the likelihood of

[30] This is why decisions taken outside the EU framework may compromise the EU's authority: even if they do not contradict EU norms, by supplementing and in some cases duplicating them with parallel institutions they imply the insufficiency of those norms to deal with concrete situations.

[31] Cf. Gross and Ní Aoláin (*Law in Times of Crisis*), p. 161.

[32] Consider the succession of market-stabilizing measures (EFSF, Euro-Plus Pact, Six-Pack, ESM, and Fiscal Compact), as well as repeated revisions to the Memoranda of Understanding.

duplication, something which makes the status of each uncertain, as well as compromising the wider intelligibility of the order as a whole.

In his study of political dictatorship, the German political scientist Ernst Fraenkel observed that formal rules and informal methods can exist in parallel, the former presenting an image of procedural order belied by the potential for recourse to the latter. A 'normative state', shaped around laws and procedures and intended for public appreciation, could be coupled with a 'prerogative state', indifferent to them and willing to override them. His term to describe this was the 'dual state'.[33] The concept is instructive for thinking about the EU after its experiences with emergency politics, even if characterizing it as a dictatorship would ill suit its dispersed structure. The Union retains on the one hand the hallmarks of a formalized system, and it may operate consistently with these in a number of domains for substantial periods. Executives may restrict themselves to practices broadly in line with the EU's division of powers and its stated commitments to democracy and accountability. At the same time, the potential for the substitution of formal structures with informal, ad hoc methods is retained. There are the seeds of something more arbitrary, 'in which the dominant officials exercise their discretionary prerogatives'.[34] Poul Kjaer aptly calls it a 'dual union'.[35] Given the prospect that the polity's norms will be adhered to quite selectively, its procedural authority remains questionable.

To be sure, separating the emergency regime from something one might call normality is not easy at the best of times. Norms are always to some degree susceptible to exceptionalism, as Schmitt and others have convincingly insisted.[36] A sharp distinction between the unbound character of exceptionalism and a period of rule-bound 'normality' to come is exactly one of the misleading implications of the exceptionalist imaginary. The contrast is overdrawn, probably in part because it is intended by those employing it to validate the idea of exceptional measures for exceptional times. There is something of the 'dual state' about all political orders. But even if constitutional reassurance can therefore never be

[33] Ernst Fraenkel, *The Dual State: A Contribution to the Theory of Dictatorship* (Oxford: Oxford University Press, 1941/1917).

[34] Fraenkel (*Dual State*, p. 3).

[35] Poul Kjaer, 'European Crises of Legally-Constituted Public Power: From the "Law of Corporatism" to the "Law of Governance"', *European Law Journal* 23 (5) (2017), p. 429; cf. Hauke Brunkhorst, 'The European Dual State: The Double Structural Transformation of the Public Sphere and the Need for Repoliticization', in Jiří Přibáň (ed.), *Self-Constitution of European Society: Beyond EU Politics, Law and Governance* (London: Routledge, 2016). Something like this dual structure of today's EU is evoked positively by van Middelaar (*Alarums and Excursions*, pp. 165ff.) as the distinction between the Community (rule-bound) and Union (discretionary).

[36] Carl Schmitt, *Dictatorship* (Cambridge: Polity, 2013 [1921]). On discretion as endemic to statecraft: Kim Lane Scheppele, 'Exceptions that Prove the Rule: Embedding Emergency Governing in Everyday Constitutional Life', in Jeffery. K. Tulis and Stephen Macedo (eds), *The Limits of Constitutional Democracy* (Princeton: Princeton University Press, 2010).

absolute, it can be present in different degrees. In the circumstances analysed it must be judged weak.

Coping Strategies

Procedural authority thus becomes a difficult foundation for executive agents to claim after a period of emergency rule. The damage is lasting. Time, of course, is perhaps a healer of sorts. If a period of relative calm ensues, in which the practices and rhetoric of emergency are less deployed, a process of consolidation may unfold. Likewise as executive power comes to be repopulated with new faces, the prospect of further rounds of exceptionalism may come to seem more remote. Perhaps one can even speak of a kind of 'crisis fatigue', which encourages citizens and others to project more stability into the system than there is, and thus to contribute to a process of normalization. The idea of a return to normality is surely enticing for many. But without change to the fundamentals, the 'posterior normal' will remain a quite misleading idea. There is little constitutional reassurance to be taken.

How can executive agents proceed in view of this fact—what ruling resources do they retain? Here I shall highlight two coping strategies, both of which may be interlaced with repeated recourse to the practices and discourse of emergency. One involves efforts to minimize dependence on authority by using methods of coercion and reward. Compliance with decision-making is sought through the application of power, intimidation, favour and uncertainty, more or less masked by institutional structures. A second involves the effort to substitute procedural authority with a different kind of authority, of which a revived authority-of-results would seem the most likely candidate.

Consider first the strategy of seeking to govern in ways that are minimally dependent on authority. Compliance is then sought through directly coercive or reward-based mechanisms rather than by appeal to normative criteria. Here one may note the important distinction between *following* norms and *conforming* to them. The first involves non-instrumental action: norms are adhered to because their demands are accepted as valid. Respecting them is the appropriate thing to do. By contrast, conforming involves action principally motivated by instrumental concerns. Demands are adhered to not because they are recognized to have normative status (this may be doubted), but because doing so brings some kind of advantage or avoidance of harm.[37] The latter opens the possibility of compliance even where the authority of the political order is weak.

[37] Brennan et al. (*Explaining Norms*), p. 218: 'We conform with a norm, in the sense we have in mind, when we act in accordance with the norm, not because of the norm, but because of other considerations associated with the norm. When an agent follows a norm, she treats the norm as a

A pattern observed at several points in the book is how discretion in the name of crisis management may be coupled with efforts to establish new binding mechanisms. To the extent that the crisis is regarded as due to errors of decision-making, the aim becomes to discipline the relevant agents so that these will not be repeated. Some such new constraints can be thought of as legal rules, while others are directed at specific agents and so lack the generality associated with law.[38] While these may have several functions, one no doubt is to strengthen decision-making by creating further reasons for compliance with its demands beyond those associated with procedural authority, or indeed authority more generally.

Consider again one set of provisions emerging from Europe's emergency politics of the 2010s—the budgetary rules of the European Semester. Given the context of crisis in which they were introduced, perhaps few relevant parties will feel it necessary to comply with these provisions *because* of some sense they are normatively valid. The public rationale for their introduction was essentially pragmatic, i.e. to avoid another Greek scenario. Yet actors have instrumental reasons to adhere to them, including the fines associated with breaking their terms or the desire to maintain goodwill, even where they are doubtful of the permanence and even-handedness of these arrangements. That the sanctions associated with breaking these provisions are not always clear means there is little basis on which to anticipate the costs of non-compliance.

Those hoping to rule on this coercive basis may count also on the possibility that innovations passed under emergency gradually come to seem less publicly objectionable. As time passes, normative standards may be reset or become blurred. Important here is the phenomenon of authoritarian drift. Amongst a wider public, one effect of a period of pronounced discretionary decision-making may be to modify the expectations with which decision-making is met. Actions which once seemed a challenge to the polity's norms become the new empirical standard against which conduct thereafter is assessed. They recalibrate the scale—of what is thought realistic to expect, perhaps even of what is thought acceptable.[39] Though ever more drastic discretionary manoeuvres may be advanced, each may appear both futile to resist (since they form a recurrent pattern) and merely an incremental advance on previous ones (since the wider shift they amount to is

(non-instrumental) reason for acting in accordance with that norm: the norm is "internalized". When an actor conforms with a norm she treats the norm as providing *instrumental* reasons to act in accordance with the norm.'

[38] Cf. Kilpatrick ('On the Rule of Law'), p. 340; Armin von Bogdandy, 'Ways to Frame the European Rule of Law: *Rechtsgemeinschaft*, Trust, Revolution, and Kantian Peace', *European Constitutional Law Review* 14 (2018), pp. 682–3; Beukers, de Witte, and Kilpatrick (*Constitutional Change*).

[39] Cf. Oren Gross, 'What "Emergency" Regime?', *Constellations* 13 (1) (2006), p.82; Gross and Ní Aoláin (*Law in Times of Crisis*), p. 236; Klieman ('Emergency Politics'), p. 1.

diffused over time). Acts of exceptionalism may be received thus less as sources of provocation than as sources of information about 'how things work' and the need to lower expectations accordingly.[40] Public concern may never reach then the acuteness that invites outright rejection, or may be diverted into a fatalist attitude. Executive powers may hope that the perpetual resetting of reference-points this involves makes it easier for citizens to live with demands whose authority they weakly believe in.

This coercive legacy of emergency rule is something we shall return to later when discussing principled disobedience. What also deserves attention however is how emergency rule can bring consequentialist forms of authority to the fore. If executives cannot plausibly disavow their discretion, the flipside is that they may claim credit for anything that passes for a positive development under their stewardship. Emergency rule puts executives, their decisions and their personas to the fore. If their leadership can somehow be seen as effective, the basis for a non-procedural form of authority emerges—buttressed perhaps by claims by the same elites to be those best placed to judge *whether* it is effective. These are patterns we shall look at in the next chapter, and are one way to read the high levels of public satisfaction recorded for some officials at the height of the Euro crisis.[41] To be sure, this form of authority carries an unpredictable and arbitrary flavour, and is a precarious basis for rule. But for a period of time it may be a source of popular support, as the wider history of emergency politics bears out.[42]

The politics of emergency is conducive to such authority also because the institutional structures that survive it, or the measures initiated under it, may come to seem valuable precisely because they have weathered a period of crisis. They have proven their durability in extreme circumstances. This is all the more true if certain provisions and actions can be presented as having helped to resolve the crisis. Note how the ECB was widely credited with restoring market confidence in the Eurozone—or simply 'saving' it—with the bond-buying policy initiated in September 2012.[43]

Importantly, the search for non-procedural authority need not rely only on consequentialist ideas but can invoke a compound of these and others. We noted in Chapter 1 the phenomenon of the 'Five Presidents'—of the Council, Commission, Eurogroup, ECB, and European Parliament—in whose name visions of the EU's future were released during one of the most intense periods of emergency

[40] Cf. von Bogdandy and Ioannidis ('Systemic Deficiency'), p. 73: 'Persons that encounter such a systemic failure of the rule of law thus modify their expectations instead of insisting on them.'

[41] The public approval for Italy's Prime Minister Mario Monti hit 52 per cent after ten months in power: see *La Repubblica*, 'Più fiducia in Monti…', 17 September 2012 (http://www.repubblica.it/politica/2012/09/17/news/sondaggio_fiducia_monti-42667390/).

[42] On the use of emergency rule as an authoritarian device: William Scheuerman, 'Survey Article: Emergency Powers and the Rule of Law After 9/11', *Journal of Political Philosophy* 14 (1) (2006), pp. 61–84.

[43] For analysis: Tooze (2018).

rule.[44] This can be seen as an effort to graft together democratic, procedural and technocratic forms of authority in conditions where each taken separately is weak. Such formations were flanked by efforts to expand the authority claims of individual institutions beyond those they were customarily associated with. Traditionally thought of primarily as technocratic, the head of the Commission started to position the institution as carrying 'political' authority too,[45] while figures in the ECB started to explore the idea that it might have a quasi-*democratic* connection to the people of the Eurozone.[46] All these can be read as efforts to find authority when there are persistent doubts about the possibility of constitutional reassurance.

Conclusion

Emergency rule presents itself as a self-contained episode, the interruption of something enduring. The very idea of exceptional measures implies the awaited resumption of political normality, with the procedural authority this enables. But it is less clear this resumption can always happen. Once executives have shown the extent of their willingness to use discretion, their challenge is to demonstrate they can step back from it. In the context of the contemporary EU, claims to this effect are difficult to make. Emergency politics is hard to consign to the past. An enduring mismatch arises between the formal rules of how the EU is supposed to work and the informal, improvised practices that can be expected of it—a version of Fraenkel's 'dual state'.

The most basic implication of the chapter is that one can expect further recourse to the transnational politics of emergency in future. Certainly, agents of established power may seek to present the institutional structures they work with as nearing consolidation, needing just a few touches here and there to 'complete' them. Such efforts at normalization may even have public success. But they rest on presenting the norms and procedures of the EU as more binding than one can plausibly treat them. The prospect of renewed extreme conditions, the enduringly weak constraints on executive discretion, and the continuity of decision-making personnel, speak against the supersession of emergency rule. Indeed, resort to it may be expected partly so as to cope with the problematic

[44] See 1 July 2015, http://europa.eu/rapid/press-release_IP-15-5240_en.htm.

[45] On Juncker's pride in having 'political heavyweights' in Commission—former PM's etc., and generally wanting to be 'more political': Desmond Dinan, 'Governance and Institutions: A More Political Commission', *Journal of Common Market Studies* 54 (2016), Annual Review, pp. 101–16; Pamela Pansardi and Francesco Battegazzorre, 'The Discursive Legitimation Strategies of the President of the Commission: A Qualitative Content Analysis of the State of the Union Addresses (SOTEU)', *Journal of European Integration* 40 (7) (2018), 853–71.

[46] Hjalte Lokdam, *Banking on Sovereignty – The Political Theory of Central Bank Independence and the European Central Bank*, PhD thesis (London: LSE, in preparation).

legacy of past episodes of the same. Once procedural authority is compromised, executives may be tempted to use further rounds of emergency politics to intimidate publics into accepting a polity whose normative foundations are weak. Constitutional reassurance seems quite out of place.

When procedural authority is persistently in doubt, governing agents must seek other bases on which to ground their actions, including but not restricted to coercion. The next chapter explores this proposition more closely by examining how the politics of emergency connects to the EU's long-standing association with authority grounded in expertise.

5
Iterations of Technocracy

One of the familiar rationales for a transnational order is to achieve certain ends more efficiently than can be done by states acting alone. The EU has often been portrayed as an exercise in technical reason: a purposive association to be valued on instrumental grounds. Its critics have often seen it in the same technocratic terms, contending that too many decisions are taken in the name of apolitical expertise rather than contestable goals and values. We may wonder then how far the principle of technocracy is compatible with the conditions of emergency rule. Does the politics of emergency mark a break with technocracy, signalling perhaps the demise of the EU's technocratic pretensions? Or is there some level of consonance between the two, perhaps even a sense in which emergency rule permits an *intensification* of technocratic claims?[1]

Reasons to expect tensions are not hard to conceive. A politics of expertise is sometimes said to prosper in conditions of stability.[2] Turbulent times spell a challenge to bureaucratic routines, and would seem to prioritize the firm hand of leadership over patient loyalty to the demands of rationality. A science of the public good is a science for the good times: when things take a turn for the worse, a more pragmatic form of governing ensues. Technocracy and emergency would in this sense be opposites, leaving the EU 'caught between technocracy and the politics of the exception'.[3]

Yet times of uncertainty can also create an appetite for answers that favours technocratic claims. Indeed, an historical perspective suggests that agents of expertise have often come to prominence exactly in conditions of upheaval. In 1930s Europe and North America, movements advocating the advance of technocracy appeared in reaction to the socio-economic challenges of the Great Depression.[4] Claims to expertise, and a measure of public appetite for them, were very much an ingredient in the economic state of emergency of the period.

[1] Cf. Chalmers et al. (*End of the Eurocrats' Dream*); Jürgen Habermas, *The Lure of Technocracy*, trans. Ciarin Cronin (Cambridge: Polity, 2015); Ignacio Sanchez-Cuenca, 'From a Deficit of Democracy to a Technocratic Order: The Postcrisis Debate on Europe', *Annual Review of Political Science* 20 (2017), esp. p. 354.

[2] Miguel Angel Centeno, 'The New Leviathan: The Dynamics and Limits of Technocracy', *Theory and Society* 22 (3) (1993), pp. 307–35.

[3] Štefan Auer, 'Merkel's Germany and the European Union: Between Emergency and the Rule of Rules', *Government and Opposition* (2019).

[4] William Akin, *Technocracy and the American Dream: The Technocrat Movement, 1900–1941* (Berkeley: U. of California Press, 1977); Antoine Picon, 'French Engineers and Social Thought, 18–20th Centuries: An Archeology of Technocratic Ideals', *History and Technology* 23 (3) (2007), pp. 197–208.

Politics of Last Resort: Governing by Emergency in the European Union. Jonathan White, Oxford University Press (2020).
DOI: 10.1093/oso/9780198791720.001.0001

Moreover, these claims were partly a form of nostalgia for the supposed rationalism of earlier periods of crisis government, in particular the efficiencies attributed to wartime planning.[5] If one goes by these historical cases, emergency and technocracy show traces of an elective affinity.

To move beyond these general impressions, we need to look more closely at the ways in which the technocratic principle of efficient, knowledge-based rule fares in relation to some more specific features of transnational emergency rule. As the present chapter shows, there are aspects of technocracy that are clearly undermined by fast, extra-institutional decision-making in the name of exceptional circumstances. Codified forms of institutional knowledge are especially disrupted. But at the same time, other types of insight that officials can claim are less destabilized, indeed bolstered. It becomes evident that there is more than one current to technocracy. The claim to knowledge that informs technocratic rule has generally been understood as a claim to scientific understanding, a claim to *know-that*. Yet in the context of emergency rule, it is another knowledge-claim that agents must rely on—knowledge as *know-how*. The chapter examines how this happens and the political implications it carries. The central contention is that, rather than signalling the termination of technocratic reason, the politics of emergency signals its renewal.

I approach technocracy as a principle of governing, not as a certain type of institution or official. Whereas the patterns of governance examined in Chapter 3 are some of the relevant institutional and behavioural forms, here our focus is on the legitimizing principle that underpins them. Institutional features such as the non-election of decision-makers, and sociological features such as the qualifications of individuals, will always be features of relevance, but they are no more than indicators of the technocratic phenomenon, and in some cases their presence or absence may be immaterial. To examine the relations between technocracy and emergency rule is to examine whether officials can plausibly rationalize their actions by reference to expertise when acting unconventionally in response to extreme circumstances.

Technocracy and the Challenge of Emergency Rule

Common to technocracy in all its forms is an appeal to expertise as the basis of legitimate rule. According to the technocratic principle, decisions should be guided by an informed understanding of the most efficient means to achieve determined goals. As one definition has it, 'technocracy is defined as a form of representation stressing the prominence of expertise in the identification and

[5] Akin (*Technocracy*), p. xiii.

implementation of objective solutions to societal problems'.[6] Though we shall encounter some important distinctions, the claim to expertise in problem-solving is a constant.

In this 'ideology of method',[7] the ends of decision-making are largely assumed. As a form of instrumental rationality, technocracy is the art of delivering rather than reflecting and choosing. It is generally not tied to a procedural conception of the good:[8] it makes no appeal to an ongoing political process to identify the goals worth pursuing, and attaches no particular value to public deliberation and opinion-formation as the measure of what should be done. Technocracy presents itself as about the efficient solution of recognized problems on which there is little reasonable disagreement.

The technocratic principle accordingly vests power in those said to wield the relevant knowledge. Technical qualifications are likely to be a prominent measure of expertise, but these are the markers rather than the principle itself.[9] Typically, the technocratic orientation is said to demand political objectivity. Technocrats may be understood to have repudiated motivations of interest and ideology: hence non-membership of parties is another conventional signifier. But again, it is the claim to expertise rather than these supporting attributes which is at the heart of the technocratic approach. We may postpone assessment of the *validity* of the expertise that officials lay claim to and the extent to which it genuinely informs their actions. While these are questions ultimately one cannot avoid, it is instructive to look first at the structure of the claims made. In the first instance, we are interested in how far actions permit rationalization—be it to publics at large or to fellow decision-makers—by appeal to the technocratic principle.

That the transnational domain has long been receptive to the technocratic principle is well established in EU scholarship.[10] It is the governing rationality one may expect of a purposive association, in which the instrumental capacity to achieve certain specified ends is the principal measure of success. A purposive conception of transnational order has often been developed in deliberate contradistinction to the partiality, conflict, and partisanship attributed to national politics. Transnational administration, certainly but not only in the context of European integration, has tended to be given connotations of the universal, the

[6] Daniele Caramani, 'Will vs Reason: The Populist and Technocratic Forms of Political Representation and Their Critique to Party Government', *American Political Science Review* 111 (1) (2017), pp. 54–67.

[7] Centeno ('New Leviathan').

[8] Christopher J. Bickerton and Carlo Invernizzi Accetti, 'Populism and Technocracy: Opposites or Complements?', *Critical Review of Social and Political Philosophy* 20 (2) (2017), pp. 186–206.

[9] One of Putnam's observations is that those who are technicians by background may not adopt the 'technocratic attitude' (Robert Putnam, 'Elite Transformation in Advanced Industrial Societies: An Empirical Assessment of the Theory of Technocracy', *Comparative Political Studies* 10 (3) (1977), pp. 383–412). One may add: those who are not technicians by background are not debarred from this outlook.

[10] Claudio Radaelli, *Technocracy in the European Union* (London: Routledge, 1999).

post-political, the consensual and the 'independent'[11]—connotations hospitable to technocratic authority, helping to rationalize the marginalization of mass publics and the weakness of parliamentary chambers. The appeal to expertise also provides intellectual support, as well as a coordinating resource, for the networked practices of governance. Technical expertise has been studied as a kind of common language facilitating the cross-national and cross-institutional communication, cooperation, and consensus-formation that governance requires.[12]

How then do claims to technocratic expertise fare in the context of emergency rule? We have identified this as a mode of governing that departs from legal-political norms and the procedures and institutions that plausibly serve these in the name of exceptional circumstances. Materially it takes the form of heightened executive activity, which in the transnational context involves multiple institutions in a precarious and shape-shifting pattern of coordination.

The transgression of legal and political norms arguably represents the least challenging aspect. The technocrat has little reason to privilege such standards as state sovereignty, interstate equality, and democratic control. In an outlook that claims supra-political expertise as its guide, and doubts the possibility of reasonable disagreement regarding the ends and means, such norms will at most be prudential concerns. Efficiency in problem-solving need depend on none of them, even if sometimes it can be better served by respecting them. Emergency rule is in this respect no particular affront. Indeed, insofar as the transgression of such norms makes it harder for decision-making to be justified by appeal to democratic or legal criteria, it consolidates a space in which technocratic claims can be advanced.[13] This is one reason why political authorities may be inclined to involve technocrats in the practices of emergency rule. When the focus is on action in the name of objective demands rather than legal or political considerations, the agents of technocracy are in their element.

When we look more closely at the rationale for emergency rule however, some tensions appear. Emergency rule involves appeal to *exceptional circumstances* as the basis for exceptional measures. It involves identifying a sudden onset of problems, leading to situations out of keeping with the world as it was before. Present and past are cast as discontinuous, and the unexpected character of the challenges arising is emphasized. Further, if one takes seriously the exceptionality of the difficulties faced, the explanations that can then be given of them are limited. Appearing suddenly and without context, they appear as something like 'acts of God' rather than the culmination of long-standing factors.

[11] Vauchez (*Democratizing Europe*); Ole Jacob Sending, 'The International Civil Servant', *International Political Sociology* 8 (3) (2014), pp. 338–40; Jonathan White, 'Responding to Norm Indeterminacy Outside the Nation-State Frame', *Comparative Sociology* 9 (5) (2010), pp. 611–30.

[12] Stephen Turner, *The Politics of Expertise* (London: Routledge, 2014), pp. 199ff.

[13] Cf. Stephen Turner, 'What is the Problem with Experts?', *Social Studies of Science* 31 (1) (2001), p. 129: 'Expertise and discretionary power are notions that are made for one another.'

Against the backdrop of exceptionalist thinking, it becomes harder for officials to tell a story about their actions being guided by scientific knowledge. An exceptional situation is one for which previous knowledge offers little instruction. Empirical data depends for its relevance on the continuity of past and present: if that continuity is openly questioned by influential figures, the possibility of empirically driven policy is questioned too. Where there are aspirations to make policy responsive to the identification of long-term regularities and causal processes, the notion of 'exceptional circumstances' will thus always be a major concession.

To suggest a set of events was unpredictable, even inexplicable, is to accept a narrative stressing the limits of what can scientifically be known. Why should problem-solving experts in this mould be granted authority if the problems of the highest order resist their capacity to foresee them? Moreover, if technocratic governing should lead to well-designed outcomes, and the same officials have been at the helm for some time, why do such problems even arise? If the expertise on offer is implicated in the making of the emergency situation, its credibility should presumably take a hit.[14] Whereas the logic of the 'shock doctrine', at least for its purveyors, is that moments of crisis offer the opportunity to bring in new expertise, things look different if the same expertise was implicated in the making of the shock.

To be sure, there are resources that technocrats in this mould can appeal to in response. They can emphasize the flip-side of the notion of exceptional circumstances—that there is a period of normality to come, when all that was thought to be known prior to the emergency period will be relevant once more. They can seek the explanation of events in the *mistakes* actors have made, which—if sufficiently rare—can perhaps be regarded as irrelevant to a science of good government. They can present the situation as a learning process and adapt their theories, or develop new paradigms of expertise.[15] They can also seek to overturn the exceptionalist view altogether and deny the uniqueness of the moment, though their power to reshape the narrative may be limited, plus to do so might underline suspicions about their own culpability in the course of events.[16]

A further tension becomes visible when one considers the modes of decision-making used in crisis management. Interventions under the sign of emergency are expected to be speedy. Decisions cannot wait for the accumulation of new evidence in the manner of scientific enquiry: they are expected to be made in the imperfect conditions of the present.[17] In the transnational context, moreover,

[14] By contrast, in the 1930s examples mentioned above, the technocratic claim was advanced by those previously marginal to decision-making.

[15] Cf. emerging notions of 'systemic risks'—things which yesterday's technocracy did not espouse.

[16] In this sense they face a basic choice: admit negligence and preserve the idea of a policy science, or 'save face' by embracing talk of exceptional circumstances, sacrificing science in the process.

[17] Cf. Turner (*Expertise*).

they require coordination between agents spread across several executive institutions. The Troika, as we have seen, expressed the centrality of cross-institutional networks. At one level, this might be seen to augment the scientific resources available for technocracy. The pooling of information between the IMF, Commission, and ECB meant each had some access to the perspectives of its peers. Each institution could offer a different kind of knowledge: the IMF had information on debtor states *as IMF members*, the Commission on such states *as Eurozone members*, and the ECB could offer information on the *Eurozone as a whole*, including on banking, finance structures, and monetary policy. Official reports from the period indicate that this information-sharing was felt to enrich the available expertise.[18] As a report by the European Court of Auditors put it, 'the variety of expertise and experience among the three institutions' staff made it possible to produce more thorough assessments'.[19] Seeing like the Troika meant seeing from several angles at once.

But the same multi-perspectival gaze of the Troika was also something that complicated the claim to scientifically informed rule. It entailed duplication of formal methodologies and metrics of evaluation, and thus the prospect of compromises between those preferred by the different institutions, sought at speed rather than as part of a gradual process of alignment.[20] Such compromises relativize the authority claims of each model and its associated expertise.[21] Reacting to such mismatches, representatives of the different institutions may be inclined to see their own measures as the true ones and their peers as moved by non-technical criteria.[22] A desire to maintain the credibility of their expertise in the face of this competition may lead to intransigence and the unwillingness

[18] See e.g. IMF ('IMF and the Crises'), p. 41; Kincaid ('IMF's Role') pp. 25–6. On the contribution to the Troika of different angles of expertise: 'ECB's replies to the questionnaire of the European Parliament supporting the own initiative report evaluating the structure, the role and operations of the 'troika' (Commission, ECB, and the IMF) actions in euro area programme countries', 2014 (https://www.ecb.europa.eu/pub/pdf/other/140110_ecb_response_troika_questionnaireen.pdf): 'The staff of the three institutions cooperate in a very good and fruitful manner. The different perspectives and experiences that the three institutions bring to the table provide for a more complete assessment and minimise possible errors or omissions.' Cf. Commission Response to European Parliamentary 'Questionnaire supporting the own initiative report evaluating the structure, the role and operations of the 'troika' (Commission, ECB and the IMF) actions in euro area programme countries', 2013: 'The three [Troika] institutions have all brought different expertise to the negotiations.'

[19] European Court of Auditors ('Financial assistance'), p. 47; p. 24 on the Commission tapping IMF experience.

[20] Cf. Henning (*Tangled Governance*), p. 34; p. 247 on how 'the institutions competed in these areas [of analysis] and the debate among them was often intense and sometimes public, even bordering on vitriolic at times.' The IMF's own 2016 report ('IMF and the Crises', p. 41) is retrospectively critical of the Troika arrangement, emphasizing the unclear division of responsibilities between Troika partners, leading to duplication.

[21] IMF ('IMF and the Crises'), pp. 41/42. Arguably this is why the German government wanted IMF involvement—see Henning (*Tangled Governance*, p.213): 'By including the Fund in the institutional mix, Berlin and the other creditors foster competition among institutions and elicit alternative analyses and recommendations—without sacrificing decision authority on debt relief.'

[22] E.g. IMF ('IMF and the Crises'), pp. vii-viii.

to abandon existing methodologies. In quieter times, such differences can be negotiated, and where they cannot be, inaction remains a generally accepted option, hence they can be overlooked. Under the sign of emergency, the pressure for immediate coordination excludes this. As a monistic outlook, technocracy is more at home with the single methodological gaze.

It is sometimes argued that crisis politics causes problems for technocracy because criteria other than efficiency become the basis of decision-making.[23] Notions of banks 'too big to fail' express the idea that there are social costs to be factored into decision-making: what the technocrat might call 'political' considerations come to the fore. Yet in principle these are susceptible to re-description as questions of efficiency, properly understood. Decisions responsive to too-big-to-fail considerations may pose a challenge to a particular policy paradigm (e.g. strands of neoliberalism), but not necessarily to technocracy itself: indeed, identifying when an economic institution is 'too big to fail' depends partly on a claim to expertise. What does pose the larger challenge is that each institutional enactment of the technocratic principle must suddenly reckon with competing versions. What otherwise could be treated as a neutral operation must now be advocated against counter-views. Method is revealed as convention. In this way each conceptualization of efficiency is put in question, and the very idea of efficiency as an apolitical, objectifiable criterion undermined.[24]

The demand for swift decision-making also encourages the circumvention of institutional procedures and the personalization of power. Chapter 1 described this as *de-institutionalization*: a pattern whereby individuals at the apex of power go outside standard operating procedures and rely on personal ties that cut across institutional divisions in order to reach faster settlements. Insofar as forms of expertise may rely on institutional procedures as part of their claims to be engaged in impersonal, scientific rule-making, again such features of emergency rule challenge this rendition of technocracy.

It seems clear then that the technocratic principle is in some ways quite at odds with executive discretion under the sign of emergency. More precisely—and to anticipate distinctions we shall return to—what comes especially under pressure is rule grounded in a certain idea of the scientific method. It is the idea of a policy science, aimed at identifying durable and objectifiable truths of the political realm by the application of institutionalized procedures of discovery, and intended to convert these into stable and rule-like prescriptions, which suffers especially in the context of emergency rule.

[23] Centeno ('New Leviathan'); cf. Pieter De Wilde and Michael Zürn, 'Can the Politicization of European Integration be Reversed?', *Journal of Common Market Studies* 50 (S1) (2012).

[24] This is the sense in which Vauchez (*Democratizing Europe*, pp. 5, 57) correctly observes a challenge during the Euro crisis to the authority of 'the institutions of "independence"'.

Emergency Rule as an Opportunity for Expertise

Technocracy as the rule of science is only one side of the story though. Intuitively one also grasps a rather different conception of expertise, in which what distinguishes the best is exactly their capacity to contend with challenging circumstances. To identify situations as exceptional and as beyond the reach of the usual governing formulas, and to claim knowledge-based authority nonetheless, is in some ways a quintessential claim to expertise. Knowing what needs to be done when standard techniques offer little guide (and equally, what should *not* be done and should *not* be tried) is an outlook fully consistent with the concept of technocracy we began with. Indeed, it might be said that the true depths of expertise can only be revealed and performed in exceptional moments such as these. It is only then that individuals can truly show insight and judgement. A certain image of expertise would seem to *rely* on ideas of the exceptional situation.

We noted above that there are historical cases when claims to technocratic expertise have been advanced in contexts consciously regarded as exceptional. Movements for rule by technocracy arose both in 1930s America and France as ways to handle the socio-economic and political consequences of the Great Depression.[25] Advanced by those at some remove from power, such movements often had a utopian flavour, intent on pursuing science to its logical conclusions. Yet it is interesting to note that such arguments for expert government were often advanced in the name not of the scientist but the *engineer*. Here was a figure associated not so much with the world of laboratory science as with a practical craft of problem-solving.[26] The engineer was conceived as someone dealing not with the pure truths of theory but the untidy and changing realities of societal order. The engineer was scientifically trained, but adjusted to an imperfect world and expected to show a facility for *judgement*. Moreover, he was expected to factor into his technocratic designs the possibility of unforeseen stresses: his was the art of designing in spare capacity for the unexpected, not just pursuing the optimal. (The paradigmatic case of course was the design of fortifications for use in war.[27])

[25] Picon ('French Engineers').

[26] See Akin (*Technocracy*) pp. 54–5 on the 'myth of the engineer' in the thought of US technocrat Rautenstrauch: 'The engineer exhibited even more unique qualities than the pure scientist. He retained the objective purity of the scientist while involved in worldly affairs. The scientist ceased being a professional when he ventured outside the laboratory, but the engineer became a professional only when he left the world of laboratory theory. He thus brought "reasoned and sound scientific judgement to the solution of the industrial problems of our national life".' Rautenstrauch's ideas and similar ones were a counterpoint to the more purely scientistic views of Howard Scott and others in the same movement (Akin, *Technocracy*, pp. 89ff.). A similar spread of views was evident in one of technocracy's forerunners, F.W. Taylor's Scientific Management: see Richard G. Olson, *Scientism and Technocracy in the Twentieth Century* (Lanham: Rowman & Littlefield, 2015), p. xi.

[27] On the historical origins of engineering as a governing rationality in France: Picon ('French Engineers').

While technocracy as the scientific method was a conception well represented in the ranks of these movements, and the one emphasized by Hayek in his critique of 'scientism',[28] it was flanked by a conception more conducive to the government of exceptional situations.

Here one catches a glimpse of a claim to expertise that does not rely solely on a science of regularities and causes, and that therefore has less reason to be discomforted by situations experienced as exceptional. It puts less emphasis on the objectification of knowledge in propositional form, with reliance on formal methodologies, metrics, and the like, emphasizing instead a more personalized form of insight and the handling and anticipation of irregular situations.

If we return for a moment to the politics of the Euro crisis, we can observe that, alongside scientific modes of expertise, there was a distinct form of expertise being invoked. The claim to know how to handle a crisis, and to have the capacity to make good on this, is one that figured prominently in the discourse of the Troika. Specifically, knowing how to restore market confidence in state finances—a problem likely to resist a formulaic approach—was a capacity emphasized. One sees a kind of *crisis expertise* being invoked by leading officials—indeed, as the constitutive idea that defines the arrangement. The standard rationale for the IMF's participation in the Troika was its capacity to provide 'programme expertise and crisis management experience'.[29] As a report by the former Director of the IMF's Office of Internal Audit and Inspection put it, the IMF's involvement was premised on its 'cross-country experience and expertise in responding to financial crises, as well as on its ability to mobilize resources quickly in emergency situations'.[30] A report on the Commission by the Court of Auditors credits it with *acquiring* this kind of expertise in the course of Troika activities, learning through doing and benefiting from its collaboration with the IMF.[31] The centrality of the ECB was likewise rationalized by one of its executive board members in terms of its being a leader in 'crisis management'.[32] That its directors should have 'broad

[28] Friedrich von Hayek, *The Counter-Revolution of Science* (New York: Free, 1955), ch. 10, pp. 94ff.

[29] IMF ('IMF and the Crises'), pp. 10, 43. NB I am interested here in how IMF participation was publicly rationalised; I do not exclude that there were reasons for involvement unrelated to expertise (cf. Henning, *Tangled Governance*, pp. 234–5).

[30] Kincaid ('IMF's Role'), pp. 6, 10–11.

[31] European Court of Auditors ('Financial assistance'), e.g. p. 47.

[32] See ECB Executive Board member Jörg Asmussen, 'Stability Guardians and Crisis Managers: Central Banking in Times of Crisis and Beyond', Goethe-Universität Frankfurt, 11 September 2012 (https://www.ecb.europa.eu/press/key/date/2012/html/sp120911.en.html), discussing the Bank's 'non-standard measures': 'why have central banks assumed such an important role in crisis management? Well, responding to the sudden financial tensions, central banks' crisis measures have contributed to calming and avoiding highly detrimental feedback loops between financial markets and the real economy. These measures needed to be implemented rapidly and central banks were in an advantageous position for acting swiftly due to their vicinity to financial markets, their operational framework and their clear and independent decision structure.' To those that say central banks may be compromised by their new prominence in crisis management, 'if paradise is lost, it may however be the right time to take on a new and a pro-active role in "the real world" or in the "new normal". We will not turn the clock backwards.'

discretion' in interpreting how to pursue the defining goal of price stability, and the option to pursue 'unconventional monetary policy measures' when deemed necessary, was claimed and in turn recognized by the CJEU.[33]

Studies of EU discourse suggest both Trichet and Draghi, as successive heads of the ECB, came increasingly to present themselves as action-oriented individuals, applying expertise to take practical steps in adversity. A shift in emphasis towards more open-ended, abstract goals is observed, as befits a discretionary attitude towards a fast-moving situation rather than a world of regularities.[34] The figure of Draghi would become the embodiment of this personalized form of crisis expertise, but it was replicated at the national level. The declaredly technocratic heads of government Monti and Papademos presented themselves in similar terms as knowing how best to respond both to the economic conditions of the day and to the technical demands of the Troika, claims supported in turn by the Troika institutions and the Council.

At the IMF, moves towards a more informal, contextual type of expertise found expression in the development of new concepts such as 'fiscal space', denoting the heuristic assessment of a particular government's ability to service its debt. Developed by senior IMF officials, the concept enabled greater discretion to be exercised in Fund evaluations, and facilitated crisis-management recommendations that were country-specific rather than based on a general template of fiscal tightening.[35] Differentiating between the capacities of smaller and larger states, it was a way of rationalizing what otherwise might look like inconsistencies of policy, and entailed a new emphasis on contextual judgement rather than scientific rules. 'Fiscal space is a judgement call', writes the author of a study, 'enabling Fund staff to weave qualitative evaluations of a government's reputation and track record into assessments of quantitative fiscal policy data.'[36] In the words of an IMF official, it is 'not a "formulaic approach", but rather "a lot of judgement involved" which is "informed by expertise, on-going experience"'.[37] Being non-quantifiable, it offered considerable latitude for interpretation, and facilitated the compromises that would be needed when reconciling IMF policy with that of its Troika partners. It also chimed with the desire of senior Fund officials to appear non-doctrinaire and flexible in the face of new challenges, in contrast to the approach of their modellers.[38]

[33] See Opinion of the Advocate General Cruz Villalon, 14 January 2015, Case C-62/14.

[34] Analysing this as a question of *charisma*: Pamela Pansardi and Pier Domenico Tortola, 'The Charismatic Leadership of the ECB Presidency: A Language-Based Analysis', *European Journal of Political Research* 58 (1) (2019), 96–116.

[35] Clift (*IMF*), pp. 17, 107–9, 212. [36] Clift (*IMF*), p. 108. [37] Clift (*IMF*), p. 109.

[38] See Clift (*IMF*), p. 8, quoting Lagarde; see also the self-descriptions of Strauss-Kahn and Blanchard. Clift notes too how IMF officials sought to lessen the emphasis on the *predictive* capacity of their expertise (*IMF*, p. 219): 'In such unusual conditions, Fund staff become all the more mindful of the limits of what predictive weight economic forecasting as a "science" can bear.' For senior Fund economists such as Blanchard, the challenge was to exercise judgement under conditions of uncertainty (Clift *IMF*, p. 19).

Each of the technocratic agents of the day was armed also of course with the codified forms of knowledge described in the previous section. But their claims to expertise involved knowing how to interpret its information reflexively and, most importantly, knowing when to set it aside in favour of a more practical understanding of the demands of crisis management. Albeit in a contemporary economic idiom, it was, we might say, the practical spirit of the engineer, and not just the rationalist viewpoint of the scientist, that was being invoked as the basis for action.

It becomes clear there is no paradox in the fact that the same technocrats who feel able to break with the existing order in exceptional times may have had a hand in its prior construction. Those who helped build the existing regime may feel better placed to set aside its rules for the sake of the *spirit* of those rules. Who knows better than the authors of the existing order the intent behind it and the elements that can be sacrificed when they no longer serve their purpose?[39] Interpreted in this more reflexive way, the technocratic principle becomes compatible with Troika officials and their partners in national ministries acting at odds with the past actions of the ECB, Commission, IMF, and Council.

One may also note how the heightened complexity of the transnational realm fosters this more practical outlook. Not only are codified forms of knowledge challenged by extra-institutional engagements under the sign of emergency, but these interactions also generate their own basis for expertise claims. Agents well-networked with other executive authorities dispersed around the Union could claim a particular kind of insight—the knowledge enjoyed by the insider, privy to decisions that others were excluded from or could see only fragments of. Secret discussions and negotiations involving the leading figures of multiple executives offered circular support for their protagonists' claims to be the only ones with true insight into the situation at hand. Having contacts in other institutions, and an understanding of the new arrangements being formed with them, became the basis of claims to be 'in the know'. Emergency decision-making in this way itself becomes the source of an expertise claim.[40]

As these reflections suggest, there is significant scope for claims to expertise in the context of emergency rule. It is exactly in situations deemed exceptional that expertise may be sought and proclaimed. How then can the same circumstances be both an opportunity for technocracy and, as suggested in the preceding section, the source of major challenges? We need to examine more closely the varieties of

[39] Cf. Luk Boltanski, *On Critique* (Cambridge: Polity, 2011), p. 147: 'To get round the rules or break them, without feeling that you have betrayed them, you have to believe, at least implicitly, that you embody, in your very person, the *spirit* of the rule. To belong to the dominant class is first of all to be convinced that you can break the *letter* of the rule without betraying its *spirit*. But this kind of belief only occurs to those who think they are able to embody the rule, for the very good reason that they make it.'

[40] Cf. Turner ('What is the Problem'), p. 139, on access to secret information.

expertise on which technocracy can be based, for it seems that the consequence of emergency rule may be to bring a different type of expertise to the fore.

Two Poles of Technocracy

We began this chapter with a standard conception of technocracy, yet less common is to observe how the technocratic principle permits quite different modes of realization. I suggest one can usefully think in terms of two ideal-types that form poles at the end of a spectrum. Gilbert Ryle once popularized a distinction between two kinds of knowledge, *know-that* and *know-how*.[41] Technocracy, I propose, as a ruling principle based on claims to knowledge, always entails an articulation of these two forms, be it closer to one or the other.

The distinction between *know-that* and *know-how* builds on a series of long-standing dichotomies, notably between theoretical and practical knowledge. At stake in these distinctions are philosophical questions to do with the relation between propositional knowledge and the knowledge expressed in practice. To use a standard example, at stake is the difference between 'knowing that' cycling requires pedalling to maintain forward motion, and 'knowing how' correctly to ride a bicycle. Ryle's aim was to show that these are two distinct forms of knowledge, neither reducible to the other, and neither dependent on some third faculty.[42] For present purposes, it is less this philosophical problematic that is of interest than the way the categories of *know-that* and *know-how* point to different ways of realizing the technocratic claim.

On the *know-that* side, there is knowledge that takes the form of statements purporting to record general truths.[43] The natural sciences are commonly viewed in these terms: the search for the laws of physics is conventionally understood as the search for universally applicable statements about objective regularities.[44] Economics and the social sciences have frequently modelled themselves in this image. The approach assumes a basic stability in the research object—it is oriented to order and perpetual truths, or at least to statements that hold true in normal times. For the same reason it may aim at *prediction*. The methods for attaining this knowledge are likely to focus on depersonalized procedures that can be replicated by peers. As a principle of governing, it is likely to suggest policy rules of long-term and general application.

[41] Gilbert Ryle, 'Knowing How and Knowing That', in *Gilbert Ryle: Collected Papers, vol. 2.* (New York: Barnes and Noble, 1949), pp. 212–25.

[42] Ryle ('Knowing How'), p. 2.

[43] A good outline is provided by Hayek in his discussion of scientism and technocracy (*Counter-Revolution*, p. 95).

[44] I leave aside whether physicists today are still committed to this model.

On the *know-how* side, one finds knowledge oriented to the handling of contextual challenges. Rather than taking propositional form, it resides in capacity, judgement, and interpretation. It is wielded by persons in their specificity, and the networks they form, rather than objectified as data, metrics, and scientific laws. It is well adapted to tackling things that change over time: in this sense it can be knowledge for exceptional times. It taps the tradition of contextual reasoning known as *casuistry*, in which judgement proceeds from the specificity of particular cases rather than the effort to subsume them under general principles.[45]

In the context of technocracy, preparedness and reactive capacity rather than prediction are likely to form its rationale. Its goals are necessarily more abstract and ill-defined. An outlook grounded in *know-how* need not entail hostility to more codified forms of knowledge, but emphasizes their need of interpretation in concrete situations, and the appropriateness of setting rigid formulations aside when the situation demands it. Rather than prizing generalizable procedures of discovery, it prizes the insight and informed intuition of the wise: for the same reason it can be revealed and adapted speedily. As a principle of governing, it is likely to prescribe recourse to discretion rather than adherence to rules.

These points are summarized in Table 5.1.

Technocracy, I suggest, rests on a blend of these two forms of knowledge-claim. It can never be centred only on *know-that*—on propositional knowledge derived through scientific procedures—as it relies on the capacity of rulers to apply knowledge to given situations. It is, after all, a mode of governing, not just of discovery. But equally, technocracy can never be centred exclusively on the supposed insight of a unique individual. Technocrats have to account for their techniques by reference to expertise that other insightful individuals could in

Table 5.1 Expertise and its application

	Know-that	Know-how
Expressed in:	Anonymous, context-independent propositions	Capacity and wisdom exercised by individuals in context
Oriented to:	Stability/normality/regularities	Change/exception
Revealed through:	Generalizable procedures, e.g. the scientific method	Insight and interpretation of individuals
Tempo:	Slow	Fast
Rationale:	Prediction	Capacity and preparedness
Outlet:	Rules	Discretion

[45] For historical overviews, and analytic arguments for rehabilitating this tradition: Albert R. Jonsen and Stephen Toulmin, *The Abuse of Casuistry: A History of Moral Reasoning* (Berkeley: University of California Press, 1988); Cass Sunstein, *Legal Reasoning and Political Conflict* (Oxford: Oxford University Press, 1996), esp. ch. 5.

principle grasp. To distinguish themselves from mere authoritarians, they must be able to account for their actions by reference to the efficiency by which they achieve sought ends. Unlike riding a bicycle, legitimate governing depends on the possibility that a rationalization of actions can in principle be given.

But even if technocracy always occupies a position between these two poles, the balance can be struck in different ways. Institutional structures can be geared more towards the accommodation of claims to *know-that* or *know-how*.[46] Specific decisions can be defended principally by appealing to the one or the other. Intra-institutional tensions and disputes may arise in accordance with these two different points of emphasis. And historical episodes can pose a challenge more to the one aspect than the other. The above-mentioned figure of the engineer, as a recurrent trope in 1930s conceptions of technocracy, represented one reconciliation of *know-that* and *know-how*, with intellectual divisions in these movements partly reflecting disagreements about how far to emphasize the virtues of the engineer over those of the scientist.[47]

Although scholarship on technocracy tends to ignore this distinction, generally focusing on the scientific (*know-that*) pole, it is in no sense foreign to how we think about technocracy. In the study of central banking, a 'rules v. discretion' debate was recurrent for much of the twentieth century and remains ongoing. It concerns how monetary policy should be determined—by fixed rules or the judgement of central bankers—and is essentially a debate between two conceptions of technocracy. Its contributors assume—mostly—that both methods of decision-making are backed by knowledge and the sincere intent to deploy it,[48] and that further criteria (e.g. public opinion) should play no role. It is a debate, in other words, recognizably *within* the technocratic outlook, and turns on the question of how efficient outcomes are best achieved: primarily by the application of *know-that* in the design of fixed rules, or primarily in the exercise of *know-how* by central bankers responding to changing circumstances.

How do these reflections connect to technocracy in the transnational sphere? The tendency I suggest has been towards a technocracy closer to the *know-that* pole. The increasing prestige of economic expertise since the 1980s, coupled with the more general concern to develop a basis for coordination that transcends national particularity, has pointed towards a technocracy of the scientific method.

[46] I differ here from the Hayekian account that locates know-how in *market* behaviour and thus external to technocracy: both forms of knowledge, I argue, are to be found in the institutional context. Cf. Friedrich von Hayek, 'The Use of Knowledge in Society', *American Economic Review* 35 (1945), pp. 519–30.

[47] Akin (*Technocracy*).

[48] E.g. Finn E. Kydland, and Edward C. Prescott, 'Rules rather than Discretion: The Inconsistency of Optimal Plans', *Journal of Political Economy* 85 (3) (1977). The original debate initiated by Friedman is more sceptical of the motivations of central bankers and thus might be construed as a debate about technocracy v. arbitrary rule (see Milton Friedman, 'A Monetary and Fiscal Framework for Economic Stabilisation', *American Economic Review* 38 (1948), pp. 245–64).

Studies of the ECB observe long-term trends towards 'scientization', as the institution's research wing expanded to develop usable economic research which in turn fed the design of rules-based policy.[49] The focus was on refining the system so it could best maintain an equilibrium, as opposed to cultivating the preparedness of officials for volatility. Price stability, a goal predicated on stable times, took precedence over readiness for extreme events. Likewise in the European Commission, advanced academic credentials such as PhDs came to be prized over the more experiential qualities of a sector-specific background, again suggesting the pre-eminence of theoretical over practical expertise.[50]

It is, one may now observe, exactly the *know-that* pole of technocracy that suffers in the context of emergency rule. It is here that its rigidities become visible and problematic. The analytical qualities associated with a science of economics (data acquisition, measurement, and the identification of causation) become increasingly vulnerable. The spotlight instead falls on a claim to knowledge oriented to situations that are irregular and demand speed of action. What are then prized especially are the *know-how* qualities of experience, practical capacity, interpretive insight, judgement, connectedness, and agility—as well as knowing what *not* to do. This is the face of technocracy which emergency rule pushes to the fore. In this perspective, speed of action is less a problem than something to be sought—indeed, the critique of the Troika arrangement from within was that at times it was too cumbersome.[51] This outlook remains recognizably technocratic in character, but adapted to the demands of emergency rule.

One reason the distinction between *know-that* and *know-how* is telling in emergency contexts is that it suggests one of the points of resilience for agents claiming technocratic authority. Ideas of *know-how* offer an alternative rationale for decision-making, and allow officials still to tell a story about themselves as enjoying a certain kind of expertise. Whereas the technocratic claim to *know-that* is always vulnerable in a crisis—why should it be accepted if it did little to ward off the problems in question?—the claim to emergency experience and know-how potentially looks ever more relevant the worse things get. Rather than a challenge to technocratic authority, upheaval becomes a resource for it. (The IMF, one may note, was further advanced down this route at the outset of the Euro crisis than the Commission or ECB: its precepts having faced difficulties in 1990s Asia and Latin America, its leading officials had already come to emphasize its credentials as an

[49] Martin Marcussen, 'Scientization of Central Banking: The Politics of A-Politicization,' in Kenneth Dyson and Marcussen (eds), *Central Banks in the Age of Euro: Europeanization, Convergence and Power* (Oxford: Oxford University Press, 2009), pp. 373–90; Stephanie Mudge and Antoine Vauchez, 'Fielding Supranational: The ECB Hyper-scientization as a Field Effect', *Sociological Review* 64 (2) 2016.

[50] Niilo Kauppi, 'Knowledge Warfare: Social Scientists as Operators of Global Governance', *International Political Sociology* 8 (3) (2014), p. 331.

[51] The IMF report ('IMF and the Crises') expresses frustration that the Troika arrangement was, if anything, too cumbersome: 'the IMF lost its characteristic agility as a crisis manager' (pp. vii–viii).

'agile crisis manager'.[52]) In this way the figures of technocracy can justify their involvement in forms of rule that would otherwise seem at odds with their claims to authority. As well as accounting for past failures, such a claim fosters a new political identity for decision-makers—as crisis managers—that stands as a resource for future initiatives, when the same agents can step forward as those best equipped to handle a difficult situation.

Secondly, the distinction helps us understand the expansion of the technocratic principle into new domains, notably into putatively *political* institutions, and the blurring of boundaries this can lead to. The logic of *know-how* facilitates the rotation of individuals across different institutions. In Europe's period of economic emergency rule in the 2010s, one saw figures generally associated with the technocratic world penetrate the world of representative democracy (Papademos was vice-president of the ECB from 2002 to 2010, before taking up office as Greek Prime Minister in November 2011; Monti was a European Commissioner from 1995 to 2004, before becoming Italian Prime Minister in November 2011). One also saw figures move from the world of representative democracy to the technocratic (Christine Lagarde was a French Minister before becoming IMF Managing Director in July 2011), as well as between technocratic institutions (Lagarde would then be nominated to head the ECB in 2019). This type of rotation—a classic theme of power-elite scholarship[53]—finds fresh impetus in the context of emergency rule given the pressure for cross-institutional coordination.

But for such rotation to work constructively for its protagonists, technocratic authority needs to be detachable from one institutional context and transferable to another. Ideas of *know-how*, more individualized and independent of institutional and analytical procedures than those of *know-that*, and more consistent with lowered expectations, are what underpin this. Individuals can be parachuted into institutional roles and acclaimed by their peers, not because they are felt to embody scientific understanding, but as a 'safe pair of hands' and a 'seasoned crisis-manager'[54]—because they can be expected to know what *not* to do, no less than their capacity to take positive action. One effect is to contribute to the erosion of the institutional boundaries between technocracy and politics.

Thirdly, the distinction illuminates the institutional divisions that can emerge and consolidate in the course of emergency rule. *Know-that* is something wielded by groups working to common templates—to shared institutional procedures for identifying expertise and bringing it to bear on policy. *Know-how* on the other hand is something wielded by individuals acting more autonomously and with

[52] IMF ('IMF and the Crises'), pp. vii–viii.

[53] Leonard Seabrooke, 'Identity Switching and Transnational Professionals', *International Political Sociology* 8 (3) (2014), pp. 335–7.

[54] See e.g. the qualities celebrated by observers in Jeroen Dijsselbloem, ex-President of the Eurogroup, when mooting him in summer 2019 as next head of the IMF: https://www.bloomberg.com/news/articles/2019-07-17/dijsselbloem-floated-for-top-imf-job-by-the-dutch-government

discretion. The distinction is likely to map onto divisions of power within an institution. Those at the apex are generally the ones best placed, and most inclined, to make claims to *know-how*, as those most directly involved in the making of consequential decisions. Those lower in the hierarchy will be more closely attached to ideas of *know-that*, separated somewhat as they are from the moment of decision, and more inclined to find the meaning of their activities in their contribution to collective procedures. By favouring the *know-how* pole of technocracy, emergency rule contributes to the detachment of those at the apex who can plausibly wield it from those tied more closely to the *know-that* strand. It serves to detach elite decision-makers from their institutions, marginalizing the latter in the process.

Towards a Precarious Technocracy of Know-how

The claimed facility of *know-how* is central, I suggest, to assuaging the tensions that arise for technocracy in difficult times. It is in the context of proclaimed exceptional circumstances that agents distributed across different executive institutions have urgent reason to coordinate, irrespective of their formal roles. Doing so puts the scientific, *know-that* logic of technocracy in question, by confronting it with its plural forms and with non-technocratic considerations. *Know-that* claims do not disappear, as they may have political utility for their masters independent of their analytical vitality: technical reasoning and quantification can be deployed rhetorically to shape the agenda, cultivating a sense of necessity for certain options and obscuring the larger stakes, even when its proponents are dubious of its scientific status.[55] But their value as a projection of expertise is curtailed. The logic of *know-how* acts to bridge these tensions. It allows those associated with institutions of expertise to work together with expressly political agents (e.g. member-state governments) without wholly compromising their technocratic status, and allows the collective as a whole to argue that its exercise of discretion is not arbitrary or merely power-political—that it is discretion backed by knowledge.

It may be tempting to see this as a story of the appearance of *charisma* in Weber's sense, i.e. a basis of authority rooted in the exceptional qualities of individuals.[56] While this usefully captures the element of personalization in

[55] Lemoine ('Measuring') pp. 15ff.

[56] I.e. 'a certain quality of an individual personality by virtue of which he is considered extraordinary and treated as endowed with supernatural, superhuman, or at least specifically exceptional powers or qualities. These are such as are not accessible to the ordinary person, but are regarded as of divine origin or as exemplary, and on the basis of them the individual concerned is treated as a "leader"' (Max Weber, 'The Nature of Charismatic Authority and its Routinization', in S. N. Eisenstadt (1968) (ed.), *Max Weber on Charisma and Institution Building: Selected Papers* (Chicago: UCP), p. 48). Weber emphasizes (p. 50) that charisma forms an *emotional* tie and is not based on the technical qualifications

play, it is also potentially misleading. In the Weberian scheme, charisma is typically understood as a point of *contrast* to rationalistic modes of rule, whereas I suggest the claim to expert know-how is *part* of the technocratic outlook. Its adherents still claim to be acting in the service of efficient outcomes, albeit under the constraints posed by exceptional circumstances, and still claim an orientation to what rationality demands. Expertise and technique, rather than personality or will, remains the basis of their claim to authority.

An important question is how sturdy and durable this alternative iteration of the technocratic principle may be. One thing it has in common with charisma is a certain brittleness. As a form of knowledge based on personal insight, *know-how* is inevitably hard to render visible and susceptible to general recognition. Technocracy of any kind depends on some degree of trust, as it is by definition an exclusive capacity only a minority can display.[57]

But whereas *know-that* is anchored in methods that can apparently be reproduced by peers, *know-how* is harder to objectify. It demands a trust for which little support can be given. As discretion comes to the fore, the notion that technocrats are simply enacting a series of specified tasks that have been delegated to them and formulated as policy rules becomes untenable—yet the basis for putting faith in their wider practical judgement is unclear. That the context in question is one of closed-doors decision-making makes the burden of trust all the greater. Furthermore, because an exercise of *know-how* puts no special emphasis on rules of general application, it is structurally similar to arbitrary rule. As with the Troika's conditionality measures, targeted at particular member-states, actions carried out in its name are always susceptible to the suspicion that they are just the expression of brute power.

Can practical know-how at least be judged by its results? Only up to a point, as appeal to it will often be made exactly in situations where things are already thought to be going astray. Its promise is to make the best of a difficult situation, not necessarily to show improvement on the present. To the extent that problems remain unresolved, there are always factors other than deficient expertise that can be invoked to explain why. This is a kind of knowledge difficult to validate by achievements.[58]

Can it be validated by authoritative others? One of the features of the transnational context is how, given the multiple sites of authority in play, agents in peer institutions can support each other's claims. Judicial actors saw fit to acknowledge the ECB's expertise in the course of the Euro crisis, underlining the need to grant it

of officials; it is 'foreign to economic considerations' (p. 52); it is also associated with the sacred rather than the profane. On it being a point of *contrast* with 'rational authority', p. 51.

[57] Turner ('What is the Problem'), p. 130.

[58] Cf. Turner ('What is the Problem'), p. 140, on the problems this presents for public evaluation.

'broad discretion' in its policy-making.[59] Without claiming economic expertise themselves, judges claimed the 'meta-expertise'[60] by which to identify those in possession of it. In this way expertise was not just proclaimed but granted formal recognition. Peer validation is only likely to resonate more widely of course to the extent that peer agents are not themselves the object of scepticism. One may wonder whether any agent of the EU sphere, including the Court, has the public authority required.

Another problem familiar from the study of charisma is that of the *reproduction* of authority. As we have seen, know-how is an expertise claim that attaches in the first instance to individuals. In the most dramatic form it involves the ECB head being lauded as 'person of the year', as 'the man who saved the Eurozone'.[61] Individuals are only in post for so long however: somehow the claim to expertise needs to be depersonalized once more. It needs to be rendered transferable, or it departs with the individuals in question.

Additional to all these structural reasons why a technocracy of know-how is precarious are of course any substantive concerns one may have about the nature of the decisions taken. Could actions in the Euro crisis really be squared with notions of expert rule? The extension of loans to states in difficulty on the financial markets was arguably a poor way to resolve the problems they faced, informed more by the interests of north-European banks than expertise in crisis management. There are reports that even Troika representatives themselves became doubtful of the policies they were involved in, feeling simply that it was too late to back out.[62] The period arguably advanced a longer-term pattern of technocratic officials becoming disenchanted with the very expertise they claim.[63]

In this sense, though they might still tell a technocratic story publicly about their participation in the process, the account's validity was ultimately dubious. The true story, one might argue, was the weak capacity of decision-makers to show genuine *know-how* when required to. Rather than displaying crisis expertise, for too long they remained attached to the precepts of existing economic science. These are the dirty secrets of the situation, however, restricted to the few and slow to emerge. It is exactly because such aspects remain hidden for some time that officials remain in a position to make claims to crisis expertise.

[59] See CJEU Advocate General's Opinion, 15 January 2015: 'the ECB must have a broad discretion when framing and implementing the EU's monetary policy, and the [national] courts must exercise a considerable degree of caution when reviewing the ECB's activity, since they lack the expertise and experience which the ECB has in this area'. The Court thereby gives concrete expression (thanks to crisis) to a principle previously only implicit.

[60] On meta-expertise: Harry Collins and Robert Evans, *Rethinking Expertise* (Chicago: UCP, 2007), p. 15.

[61] Cf. the *Financial Times* on Draghi, 13 December 2012: http://www.ft.com/cms/s/0/8fca75b8-4535-11e2-838f-00144feabdc0.html#axzz33Tb2ilre

[62] Varoufakis (*Adults in the Room*), p. 21.

[63] Cf. David Kennedy, *A World of Struggle: How Power, Law and Expertise Shape Global Political Economy* (Princeton: Princeton University Press, 2016) p. 151.

That these claims are precarious is one reason why passages of emergency rule may be followed in the longer run by efforts to reduce reliance on the know-how of individuals. Renewed notions of a technocracy of scientific *know-that* are evidently one form this can take. The ideal of a well-designed system that comes close to running itself, machine-like and quasi-automatic in its application, then comes to be rehabilitated. Alternatively, in line with the rise of algorithmic forms of decision-making, such a system might be based not on causal analysis but the correlations revealed in big data. The latter would be better seen as a form of 'post-technocracy', neither grounded in the search for scientific truths nor the judgement of skilled interpreters. Both such tendencies would seek to minimize the role of know-how, and in turn would leave agents ill-equipped to display it.

Perhaps just as likely though is that those who come to understand themselves as crisis managers will seek out new opportunities to act as such. Those who define themselves by crisis expertise will naturally seek outlets for their capacities. Especially if officials see potential here to bolster their support amongst citizens, they may seek opportunities to reaffirm these qualities.[64] An expanding array of situations would then be approached as exceptional situations demanding the application of emergency know-how. One could expect the temporal range of such expertise to be extended into the future, in the form of the anticipation of breakdowns to come. Frequent warnings about problems on the horizon allow policy-makers to cultivate a constituency for their crisis expertise, while also ensuring they are not seen to be taken by surprise next time around. Predicting bad things to come is a logical way for policy-makers to cover their backs.[65] It is also a way to galvanize others to get ready. 'Preparatory emergency management'[66] is arguably how claims to emergency know-how become embedded in policy-making for the long term. Such practices have been identified as increasingly central to the activities of the European Commission and to the proposals it puts to the Council. Here again one sees the potential for practices of emergency rule to reproduce themselves.

At any rate, the more immediate legacy of emergency rule would seem to be a recalibration of expertise away from a science of policy towards the capacity to cope with the unexpected. Such ideas of *know-how* offer a way for avowedly technocratic agents to justify their involvement in the practices of emergency rule, as well as their influence in shaping the order that follows.

[64] See Rhinard ('Crisification'), who observes based on interviews that 'most policymakers, too, believe effective crisis management is a "way out" of the EU's public support problems'.

[65] For the views of a former central banker: Howard Davies, *The Guardian*, 6 December 2018: https://www.theguardian.com/business/2018/dec/06/central-bank-warnings-are-getting-louder-and-more-frequent-howard-davies

[66] Curley ('Models of Emergency'). On bureaucratic forms: Jacqueline Best, 'Technocratic Exceptionalism: Monetary Policy and the Fear of Democracy', *International Political Sociology* 12 (4) (2018), pp. 328–45.

Conclusion

One tends to read crisis politics as something that discredits technocratic claims to expertise, or at least as an interruption in a larger story of technocratic empowerment. True enough, emergency rule is a major problem for notions of policy as underpinned by scientific expertise. In a transnational context, where such claims to expertise have long been central to legitimizing efforts, the stakes are especially high. But if one accepts that technocracy is a spectrum that spans two poles, the tensions are less acute. The elasticity of claims to practical *know-how* allows the agents of emergency rule to draw on such ideas for rationalizing exceptional measures for exceptional times. Inherently independent of established institutional templates, crisis *know-how* offers a rationale for decision-making in times of emergency when ideas of *know-that* are vulnerable. It provides intellectual cover at those moments when those exercising power are most visible. In this way a certain idea of technocracy finds outlet in the emergency context—indeed, exceptionalism may be pursued exactly to bolster claims to expertise.

Yet such appeals to *know-how* are precarious. Whenever executive power rests on a claim to exclusive capacity, it offers little basis for a wider public to identify with it, since those inside the charmed circle are inevitably positioned as different in kind. Decision-makers must be taken on trust, and where trust is absent they are easily denounced as a different breed. Moreover, presenting themselves as responding to the dictates of knowledge and the necessities of circumstance, they continue the tendency to separate decision-making from popular concerns and aspirations to political agency. Technocracy of all kinds remains a brittle approach to governing, never more so than when reliant on notions of crisis expertise.

6
Populism as the Promise of Agency

In the early 2000s, researching my first book, I interviewed groups of taxi-drivers in various regional cities of Britain, Germany, and the Czech Republic to get a sense of their views on political Europe. What might these people, away from the centres of political and economic power, have to say about the problems of public life, the workings of institutions, and by extension the EU?

If there was one overriding theme to emerge, it was a sense of impotence before socio-economic forces. With an air of fatalism, though not with indifference, participants evoked a world of intractable problems, of politicians unable or unwilling to make a difference, and of institutions that were symbols of weakness rather than sources of address, detached from ordinary concerns. Such views were most pronounced on matters economic, and the EU stood as their emblem. As a driver called Luke in the Welsh town of Swansea recalled, combining such themes, 'I picked a guy up at the station . . . he works for the government and he's working on the Euro at the moment, and he has said that we will definitely be in the Euro. Definitely be in the Euro within five years, without a shadow of a doubt. He said, if it goes to a referendum, if they say no, we *will* be going into the Euro, everything will be fixed. "We will be in the Euro", he said, "I'm working on it now."'[1]

Such sentiments of impotence in the face of the inevitable are a widespread trope of the age.[2] In Europe and beyond, material constraints on political authority, combined with narratives of globalization and the primacy of the market, point to a world of weak political agency. From the mid-1990s especially, governments came increasingly to emphasize the limits to what politics could achieve on traditionally salient questions, calling on individuals to adapt to change rather than seek to author it.[3] Nowhere were such ideas stronger than in connection with the EU, but they were hardly exclusive to the transnational sphere. Left parties

[1] Jonathan White, *Political Allegiance after European Integration* (Basingstoke: Palgrave Macmillan, 2011), p. 183.

[2] A report on perceptions of the economy in Britain likewise concludes fatalism before 'mysterious market forces' is a central theme: on 'the central challenge' as 'the belief that the economy cannot be controlled or changed by collective action': NEON, NEF, FrameWorks Institute, PIRC, *Framing the Economy: How to Win the Case for a Better System* (London, 2018), esp. p. 37.

[3] Colin Hay, 'The Normalizing Role of Rationalist Assumptions in the Institutional Embedding of Neoliberalism', *Economy and Society* 33 (4) (2004), pp. 500–27; Colin Hay and Ben Rosamond, 'Globalisation, European Integration and the Discursive Construction of Economic Imperatives', *Journal of European Public Policy* 9 (2) (2002), pp. 147–67; Matthew Watson and Colin Hay, 'The Discourse of Globalization and the Logic of No Alternative: Rendering the Contingent Necessary in the Political Economy of New Labour', *Policy and Politics* 30 (4) (2004), pp. 289–305.

Politics of Last Resort: Governing by Emergency in the European Union. Jonathan White, Oxford University Press (2020). © Jonathan White.
DOI: 10.1093/oso/9780198791720.001.0001

across Europe, to whom one might once have looked for confident assertions of mastery, came to emphasize instead the importance of adapting to the market.[4] The disavowal of political agency, and its practical fragmentation across private and technocratic institutions, has been central to the spirit of the times.

The EU's government-by-emergency in the 2010s marked, this chapter argues, an extension and intensification of these denials of agency. Transnational decision-making under the sign of emergency is characterized by heightened appeals to necessity. For the sake of warding off urgent threats, exceptional measures are demanded, and are said to be needed without delay: there is no option to wait and see. These conditions of emergency rule expand the repertoire of arguments from necessity that executive powers can give. They scale up the magnitude of the decisions to which they are applied, and also bring to the fore a constellation of political and technocratic figures united by their insistence on the need to take unpalatable measures and to rule out alternatives.

This type of governing rationality can be highly effective in passing measures that might otherwise meet resistance. As analyses of the neoliberal 'shock doctrine' bring home,[5] government by emergency can buy the compliance of those who might contest such initiatives if discussed on their merits. But it can also prompt a countervailing impulse. The concept of *populism* has been widely used in recent years to describe mobilizations against the political establishment. As I argue in the chapter's middle section, a central motif of the politics thereby described is exactly the rejection of discourses of necessity. What goes by the name of populism tends to be something presenting itself as a politics of action and volition—of recognizing problems where they exist and of tackling them as one sees fit. A world of forces that resist positive human influence is one that leaves political authority without a rationale: 'populism' claims to correct this. Its exponents announce themselves as those who will take control. While the pledge may often be a false one, it is central to the claims advanced.

This promise of agency is neither historically nor conceptually unique to populism. Indeed, one of the shortcomings of populism as a concept is how it diverts attention to related but rather different features of the currents of opinion in question. We shall examine these issues more closely. But provisionally the suggestion is that the promise of agency is at the heart of the populist appeal: it becomes one of its central mobilizing features in the wider context of a disavowal of agency by other leading actors of the public sphere. What populism ultimately amounts to in this context, to varying degrees, is a form of voluntarism.

The prospect of appeals to agency in turn can inspire renewed efforts by established executive powers to constrain it—if not for themselves then for others. As the chapter's final section explores, the pursuit of rules and coercive

[4] Mudge (*Leftism*), p. 10. [5] Klein (*Shock Doctrine*).

mechanisms, and of automaticity in their application, is a typical legacy of emergency rule. Such moves are efforts to efface the promise of agency.

Governing by the Principle of Necessity

In the political imaginary of emergency rule, appeals to necessity have a central role. Citing exceptional circumstances, officials claim that departures from norms are scarcely to be avoided and that power must be vested in the executive. Whether one sees these as moves shaped mainly by the pressure of events, or as manoeuvres with a touch of opportunism, it is clear that notions of inevitability figure heavily in their rationalization. To recall just one example, when Mario Monti presented his government's 'Save Italy' decree in December 2011, for which his EU partners had pressured to have him installed just weeks before, he declared that 'the reduction of our public debt is a vital necessity and every deviation risks plunging our country into an abyss'.[6]

Ideas of necessity often take socio-economic conditions as their reference-point. In the discourse of the Euro crisis, constraints imposed by dwindling market confidence were central. In this perspective, external forces act directly on decision-makers and compel them to adopt certain measures to maintain the existence of valued ends, be they specific policy regimes or general aspects of the socio-economic order. In line with the logic of securitization, governing occurs in the shadow of threats and must do what is required to ward these off. The principle of necessity responds then to forces external to the political realm.

Ideas of necessity can also be elaborated in a more mediated way, evoking the constraints imposed by other executive agents as they respond to the challenges at hand. This is a possibility afforded by the transnational realm especially. For any one agent, the options available are naturally restricted by the decisions of others. Whether or not sovereignty at the national level is retained in the last instance, the daily exercise of power involves coordination between many. This basic truth is easily raised to the status of a determining constraint, and acts as a resource for those wanting to defend their decisions. The interplay between EU, national and international institutions in the context of emergency rule offers considerable scope for this. The officials of the Troika commonly presented their actions as necessitated by their peers, while national executives invoked the spectre of the Troika to press for measures their national parliaments might otherwise resist.[7] Genuine constraints were in this way amplified to become demands permitting only one response.

[6] 5 December 2011, https://st.ilsole24ore.com/art/notizie/2011-12-05/monti-fuori-euro-baratro-161011_PRN.shtml

[7] See Moury and Standring ('"Going beyond the Troika"') on Portugal.

These two strands of argument from necessity are often linked. To the extent that the diffusion of power in the transnational context *does* pose genuine constraints on executive authority, agents wanting to galvanize others to a certain course of action have natural reason to invoke pressures arising in the socio-economic domain as demands that cannot be ignored, in the hope that this can unite a dispersed set of executive agents around a common decision. In a fragmented institutional field, appealing to the force of necessity may often seem the easiest way to get things done.[8] Non-hierarchical decision-making contexts that rely on the cooperation of agents dispersed across multiple institutions encourage a discourse centred on constraints.

Whether by appeal to socio-economic or political constraints, the effect is the *disavowal of agency*. In both cases, forces external to the office-holders in question are evoked to indicate sharp constraints. Certainly, executives acknowledge their role as enactors—we are not dealing with outright denial. One of their aims, after all, is typically to convince their addressees—whether populations, the markets or other institutions—that they are acting to deal with the situation at hand. What we may call their *doings*—the sheer enactment of measures responding to events—is something executive powers may be keen to proclaim. And they may also add sometimes that the actions they take are desirable in their consequences—that they are not only necessary but good. But to express agency is not just to accept authorship and to assert the desirability of outcomes but to say the actions were freely chosen. It is to say that other options—including perhaps inaction—were available, and hence that what was done was done of volition.

It is in exactly this sense that executive agents acting in the name of emergency deny their agency. However much they may play up their *doing*, and however much this may include high-profile unconventional moves, their actions are generally presented as nothing to do with volition. Though one may suspect that more choice is being exercised than is advertised, and that agency can still fairly be attributed, the public justification is all about compulsion. This is true even when agents refer to their own 'discretion', as in the well-known case of the ECB. This is discretion in the service of 'whatever it takes', not discretion as responding to the agent's freely chosen goals.[9]

One reason to emphasize this is because existing scholarship on states of exception, generally focused less on the justifications invoked than the material breaches and suspensions of legality, tends to read emergency rule more as a demonstration of agency than the context of its abjuration.[10] If our interests lie with the political and democratic implications of emergency rule, we need to

[8] White ('Emergency Europe'), p. 312; Astrid Seville, *"There Is No Alternative": Politik zwischen Demokratie und Sachzwang* (Frankfurt am Main: Campus, 2017), pp. 276ff., 417–18.

[9] See famously ECB President Mario Draghi, London, 26 July 2012: https://www.ecb.europa.eu/press/key/date/2012/html/sp120726.en.html

[10] E.g. Rubenstein ('Emergency Claims').

distinguish between the frenetic activity it gives rise to (*doing*) and the willingness of authorities to claim genuine ownership of this (*agency*).

The disavowal of political agency is a long-term tendency by no means exclusive to emergency decision-making. In the European context it can be traced back well before the 2010s.[11] In addition to the incentive provided by the EU's structure, the prominence of the technocratic principle is clearly significant. Technocracy entails an emphasis on selecting the right course rather than choosing between viable options.[12] Whether in the guise of experts responding to the dictates of scientific reason—as central bankers have tended to cast themselves—or more recently as crisis managers bringing emergency know-how to a grave situation, the non-majoritarian agencies of the EU generally present themselves as doing what has to be done. However powerful and discretionary they may be, such figures tend to avoid the suggestion they are acting of volition, with the added burdens of accountability this would demand. Instead they focus on getting ahead with what needs to be done—much like the technocrat in the back of Luke's cab.

One should also recall the long-term influence of certain ideologies on European and EU politics. Though Ordoliberalism's influence today is debated, its historical significance for European integration is perhaps less so.[13] Noteworthy for our purposes is the commitment to binding political agents to a set of foundational economic rules known as the 'economic constitution'. The Ordoliberal claim is that political agency, once it has been exercised in the decision for a certain economic model, can largely be consigned to the past. The present is a time for technical adjustments geared to maintaining the system in the face of adversity.

The influence in more recent decades of economic ideas associated with Chicago School neoliberalism marked on the one hand an expansion of the intended range of political *doing*—institutions were expected to devote themselves to positively transforming the economies of European states by such mechanisms as competition policy—coupled on the other hand with an increase in the disavowal of political *agency*—as ever greater emphasis was publicly placed on the self-organizing capacities of markets and the need to respond to their demands. Reliance on institutions to structure economic behaviour if anything increased, but in tandem with an escalation in the rhetoric of the free market as the horizon of possibility.[14] As Wendy Brown puts it, 'in the neoliberal political

[11] See Chantal Mouffe, *On the Political* (London: Routledge, 2005).

[12] Centeno ('New Leviathan').

[13] Hien and Joerges (*Ordo-Liberalism*); Bonefeld (*Strong State*); Gerber ('Constitutionalizing'); Slobodian (*Globalists*); Chapter 2, this volume.

[14] Cf. Ian Bruff, 'The Rise of Authoritarian Neoliberalism', *Rethinking Marxism: A Journal of Economics, Culture and Society* 26 (1) (2014); William Davies, *The Limits of Neoliberalism: Authority, Sovereignty and the Logic of Competition* (London: Sage, 2014).

imaginary…we are no longer creatures of moral autonomy, freedom, or equality. We no longer choose our ends or the means to them.'[15] Or as Colin Hay wrote of neoliberalism under Britain's New Labour: 'the justification for policy is presented not in its own terms, but as a necessary accommodation to the "harsh realities" of new economic times in a (superficially) dispassionate, almost technocratic manner. Appeal is…made to processes beyond the control of political actors which must simply be accommodated and hence to a dull logic of economic compulsion which is non-negotiable.'[16] As many others have documented, narratives of There-Is-No-Alternative (TINA) have been recurrent in the western world since the 1980s. Though often with variations across time and place—TINA-doctrines under Thatcher and Blair were not the same,[17] and there were also important nuances within each[18]—the same broad theme has recurred. The effect on the public imagination, it would seem, was to evoke a world of large, irresistible socio-economic forces, combined with fatalism concerning the extent to which one could do other than adapt.[19] Especially the transnational domain, coextensive with these macro forces yet seemingly powerless to redirect them, became synonymous with the out-of-control.

What then does emergency rule, of the kind applied in both economic and migration policy in recent EU history, add to this? In what sense does one see an intensification and extension of these longer-term trends? One relevant dimension concerns the rationalization of actions: the appeal to necessity becomes sharper. Policy-makers find themselves emphasizing what *cannot* be done in present conditions without further undermining a precarious situation, and highlighting the need to adopt urgently a certain response. In the standard technocratic outlook, certain courses of action demand to be followed because they are superior. Scientific, *know-that* reasoning points to an optimal set of policies, and the sooner they are adopted the better. But the possibility of delay, though undesirable, can be entertained. Rather than an existential problem, one simply risks 'falling behind'—to use a phrase commonly used in neoliberal competition policy. Things are different once one enters the logic of emergency. Political and technocratic authorities carve out a window of time in which action must be taken—that between what I have called the *anterior* and *posterior normal*. There is no longer a continuous temporal axis whereby action occurs 'the sooner the better'; rather, it is the dichotomous idea of 'now or never'.[20] To wait would be

[15] Brown (*Undoing the Demos*), pp. 41–2.

[16] Hay ('Normalizing Role'), p. 519.

[17] Seville (*Politik*), building on Hay ('Normalizing Role').

[18] On three strands of argument from economic necessity in New Labour: Watson and Hay ('Discourse of Globalization').

[19] White (*Political Allegiance*).

[20] A similar distinction is implied by Seville (*Politik*), who contrasts 'TINA' discourses of the Thatcher and Third-Way eras with those of more recent crisis management by charting a shift from '"one right way" politics, to "one possible way" politics'.

to miss the window of opportunity and potentially to be left with nothing worth defending.

This is an important difference. When actions are needed 'the sooner the better', there remains the possibility of deferring them for prudential reasons. Such concerns as public opinion, social stability, or basic questions of justice, though they will not typically be raised by technocrats as reasons for action, can still be summoned as reasons to wait by those wishing to oppose certain actions. The decision to go ahead retains therefore the character of a choice (if nothing else of timing), and those in charge will often want to claim credit for their willingness to take 'hard decisions'. When, by contrast, actions are said to be needed 'now or never', there can be no possibility to delay, and no appeal to cross-cutting considerations as reasons for caution or moderation. They are non-negotiable actions, and the figure of executive authority is reduced to something like a conduit for forces beyond its control. Rationalizing executive action in such circumstances tends to involve tracing problems to quasi-natural forces that must be adapted to rather than resisted ('pressure' from the markets, 'flows' of migration). The disavowal of agency is more absolute, in other words, as it concerns not just the substance of action but its timing.

Emergency rule is distinctive in a second respect, in that necessity is invoked not only in a more absolute fashion but is applied to decisions with the potential to be especially provocative. If 'exceptional circumstances' spell the need for exceptional measures, then the decisions for which necessity is claimed may be quite unconventional in make-up. There can be little pretence that officials are taking low-stakes, business-as-usual decisions. On the contrary, that they are taking significant measures is something they will typically want to emphasize, to convince relevant audiences that action is being taken. Such moves bring to public attention the considerable lengths to which executive agents are willing to go in the service of necessity. Exactly the norm-transgressing character of their actions serves to indicate how the demands of necessity trump countervailing considerations, in a way that TINA-style claims *within* the normal political process do not. Moreover, because such measures may be aimed not only at overcoming an emergency but preventing its return, they may be specifically designed to be irreversible. Ideas of necessity are thus asked to bear a heavier load, and to carry it forward for an indeterminate period of time.

The third respect in which emergency rule marks an advance on more general patterns of agency-disavowal concerns the structure of the agents making these claims. Informal and hasty coordination across multiple sites of executive power, supranational, national, and sub-national, is one of the hallmarks of transnational emergency rule. In place of the demarcation and differentiation of roles, one sees an opaque constellation of figures interacting in ad hoc ways—tendencies towards 'de-institutionalization', as I have called them. Several points should be noted.

One, it becomes easier for executive agents to present their actions as at the behest of others (e.g. to shift blame), since the decision-making processes become especially hard for outsiders to decipher. The observer cannot rely on formal allocations of power, such as the Community Method, as a guide to who, if anyone, has the capacity to exert agency. The shape-shifting structure of decision-making, and the discontinuity of the political entities involved (think Troika) interrupt the effort to attribute agency. However weakly intelligible the EU may be at the best of times, it is emphatically more so in these circumstances of structural fluidity.

Emergency contexts are also those when conflicts between agents are most likely to be concealed. A general preference for secrecy of decision-making is characteristic of crisis management, as agents may fear worsening the situation by openly acknowledging undesirable eventualities.[21] But especially *disagreements* are likely to be hidden, partly because they may weaken the appearance of resolve needed to handle the situation, and partly because they may raise questions about the suitability of the measures proposed. Concealing conflicts is, in part, about protecting the force of the claim to necessity as the basis of decision-making—just as, in turn, emphasizing necessity serves to further remove disagreement from sight.[22] The effect is to reinforce the perception that choice is absent from the process.

Emergency rule juxtaposes executive agents just when they are most prone to disavow their agency, and by juxtaposing them it presents them as a chain of replicas. As figures line up to pronounce the need for far-reaching actions—and the one unambiguous feature of this nexus is the expansive set of executives involved—their uniformity of message is underlined. At least as seen from afar, the Council, Troika, and executives of states are observed to be singing largely the same tune, emphasizing the need to restore stability and suppressing political and ideological differences. Institutional distinctions recede: domestic and transnational elites start to look continuous with one another, while technocratic and political elites become harder to distinguish, as all present themselves as engaged in the same fire-fighting operation. The conditions become ripe for the identification of an encompassing political class whose defining feature is its willingness to coordinate, even collude, in the service of necessity. Political authority in its entirety becomes synonymous with agency disavowed, and the terrain is ready for others to step forward with a politics that celebrates the opposite of necessity: volition.

[21] Christian Kreuder-Sonnen, 'Political Secrecy in Europe: Crisis Management and Crisis Exploitation', *West European Politics* 14 (4) (2018), pp. 958–80.

[22] Cf. Seville (*Politik*), p. 332, on Merkel et al., emphasizing in 2011 the importance of unity of message.

'Populism' as the Promise of Agency

'Take back control' was the message of the 2016 Brexit referendum. Against a background sentiment of political powerlessness, augmented from 2008 by austerity programmes administered as necessary responses to crisis, who in retrospect could be surprised? In a world experienced as one of out-of-control forces, all the more irresistible at the transnational level, one can readily understand the attraction of the promise to reassert control. With its focus on a series of 'doable' steps—diverting funds to the NHS, reinforcing borders—the Leave campaign cast itself as the only political group willing to act decisively, while others stressed only what one could ill afford to do. It set itself explicitly against the ideas of impending threat and emergency it identified in the Remain campaign—against 'Project Fear', as it came to be known. Over the following years, similar claims would be made in favour of 'Hard Brexit' over 'Soft Brexit', 'No Deal' over 'May's Deal'. The Brexit Party would take voluntarism to an extreme, discounting warnings of economic hardship while defending the country's departure from the Union 'at any cost'.[23]

Britain's decision to exit the EU drew on arguments that in part were distinctively local, but they were also expressions of a wider tendency found across the EU. If transnational emergency rule represents the intensification of governing by the principle of necessity, reactions against it commonly take the form of reassertions of political agency and will. This, I suggest, is one way to read the currents of 'populism' that became pronounced in many European states in the 2010s. What goes by this name is unashamedly a politics of volition, and as such a direct rejoinder to the disavowal of political agency.[24]

The scholarship on populism tends to home in on a number of key orientations. If we leave aside polemical uses of the term, referring simply to a politics deemed irresponsible, then typically it is used to describe a combination of features including an anti-elitist stance, an insistence on the category of 'the people', and an anti-pluralist assertion that the good of the people is incontestable.[25]

[23] Francis Ryan, 'The suffering caused by austerity helped fuel Brexit – and will only get worse', *The Guardian*, 23 May 2019 (https://www.theguardian.com/commentisfree/2019/may/23/austerity-brexit-suffering-eu-anger).

[24] For a similar reading: Sara Hobolt and James Tilley, 'Fleeing the Centre: The Rise of Challenger Parties in the Aftermath of the Euro Crisis', *West European Politics*, 39 (5) (2016), pp. 971–91; cf. Kenneth M. Roberts, 'Populism and Political Parties', in Rovira Kaltwasser, Taggart, Ochoa Espejo, and Ostiguy (eds), *The Oxford Handbook of Populism* (Oxford: Oxford University Press, 2017), esp. p. 295; Rubén Ruiz-Rufino and Sonia Alonso, 'Democracy without Choice: Citizens' Perceptions of Government Autonomy during the Eurozone Crisis', *European Journal of Political Research* 56 (2017), pp. 320–45.

[25] See *i.a.* Jan-Werner Müller, *What is Populism?* (Philadelphia: University of Pennsylvania Press, 2016); Cristobal Rovira Kaltwasser, Paul Taggart, Paulina Ochoa Espejo, and Pierre Ostiguy, 'Populism: An Overview of the Concept and the State of the Art', in Rovira Kaltwasser, Taggart, Ochoa Espejo, and Ostiguy (eds), *The Oxford Handbook of Populism* (Oxford: Oxford University Press, 2017); Cas Mudde and Cristobal Rovira Kaltwasser (2017), *Populism: A Very Short*

Additional features such as a hostility to political procedures, and a certain definition of where the boundaries of the people lie, also tend to figure prominently. Some observers emphasize that these features are logically interconnected and form a minimally coherent outlook; others observe the political 'style' that goes with them, adopted selectively by political actors according to their changing circumstances and agendas.[26]

Some of the dangers associated with clustering widely divergent political movements under the single heading 'populism' are well documented. That we may be dealing here with a concept that conflates important differences of left and right, and that has even been developed with this very intention, is one of the more powerful charges.[27] I am reluctant to endorse populism as a category of analysis, though I shall not cast it off entirely. What I want to do instead is reflect on what forms the self-understanding, and inspires some of the support for, the political groupings commonly identified as such. A key question here is whether the same features typically identified as definitional of populism by scholars working with the category are in fact the principal features of relevance. This is an assumption commonly made, with voters said to be attracted by resentment of the establishment, the seductive evocation of a primal community, or the certainties of ethical monism. There is something to be said for such readings, and we shall return to elements of them shortly.

But in the context of a political environment characterized by the widespread denial of agency, one of the features that makes so-called populist parties stand out is a secondary one neither historically nor conceptually unique to populism. Evocations of 'the people' and popular will are a way of repudiating establishment notions of necessity and presenting a collective subject guided by its own volition, willing to decide its own hierarchy of values and its own ideas of what should be done. What 'populist' forces offer, first and foremost, is the promise of agency.[28]

Given that the history of political appeals in the name of the people is bound up in the history of popular sovereignty, the centrality of the question of agency is intuitive. Without the idea that the basis for political authority rests ultimately in the people, the symbolic force of mobilizations in the people's name would be hard

Introduction (Oxford: Oxford University Press, 2017); Rogers Brubaker, 'Why Populism?', *Theory and Society* 46 (5) (2017), pp. 357–85.

[26] Contrast Müller ('What is Populism?') and Benjamin Moffitt, *The Global Rise of Populism: Performance, Political Style, and Representation* (Stanford: Stanford University Press, 2016).

[27] Marco D'Eramo, 'Populism and the New Oligarchy', *New Left Review* 82 (July/August, 2013).

[28] A methodological challenge facing efforts to identify the persuasive force of political messages is that any given statement combines multiple themes: their appeal is over-determined. In this chapter I do not exclude competing interpretations of where the motivational force of 'populism' lies, but seek to show a coherent account can be given that attaches less weight to the usual suspects (e.g. xenophobic identity politics).

to grasp.[29] Yet the contemporary analysis of populism tends to be unsatisfactory here in one of two ways. On the one hand, the basic claim to assert agency is often overlooked, with emphasis falling instead on the familiar touchstones—anti-elitism, exclusionism, etc.—with the result that something crucial is forgotten. On the other hand, no less problematically, some go so far as to build the very idea of acting on the people's will into their concept of populism, implying thereby that a reference-point shared by a much wider field of democratic ideologies is unique to just one.[30] My suggestion, to reiterate, is the following: that the promise of agency is *not* conceptually distinctive to something called populism, yet is central to the contemporary appeal of the parties we tend to class as such.[31]

Let us explore this with some illustration. Consider this statement of Marine Le Pen, leader of what was then called the *Front National*, interpreting the 2016 Brexit referendum:

> British voters understood that behind prognostications about the pound's exchange rate and behind the debates of financial experts, only one question, at once simple and fundamental, was being asked: Do we want an undemocratic authority ruling our lives, or would we rather regain control over our destiny? Brexit is, above all, a political issue. It's about the free choice of a people deciding to govern itself. Even when it is touted by all the propaganda in the world, a cage remains a cage, and a cage is unbearable to a human being in love with freedom. The European Union has become a prison of peoples. Each of the 28 countries that constitute it has slowly lost its democratic prerogatives to commissions and councils with no popular mandate. Every nation in the union has had to apply laws it did not want for itself. Member nations no longer determine their own budgets. They are called upon to open their borders against their will.[32]

Evident in this short passage are of course some of the themes typically emphasized in studies of populism—appeal to the authority of a 'people', unified in its

[29] On popular sovereignty and populism: Duncan Kelly, 'Populism and the History of Popular Sovereignty', in Rovira Kaltwasser, Taggart, Ochoa Espejo, and Ostiguy (*Oxford Handbook*); Stefan Rummens, 'Populism as a Threat to Liberal Democracy', in Rovira Kaltwasser, Taggart, Ochoa Espejo, and Ostiguy (eds.) (*Oxford Handbook*); Margaret Canovan, *The People* (Cambridge: Polity, 2005).

[30] See e.g. Cas Mudde, 'The Populist Zeitgeist,' *Government and Opposition*, 39 (3) (2004), pp. 541–63; for a critical assessment: Paulina Ochoa Espejo, 'Populism and the Idea of the People', in Rovira Kaltwasser, Taggart, Ochoa Espejo, and Ostiguy (eds.) (*Oxford Handbook*), p. 623. Preferable, because more focused, is Brubaker's suggestion ('Why populism?', p.364) that, in addition to the familiar features, populism is characterized by 'antagonistic repoliticization: the claim to reassert democratic political control over domains of life that are seen, plausibly enough, as having been depoliticized and de-democratized'.

[31] For a like-minded reading, minus a focus on emergency politics: Rummens ('Populism as a Threat'), pp. 566ff.

[32] Marine Le Pen (2016), 'After Brexit, the People's Spring Is Inevitable', *New York Times*, 28 June 2016: https://www.nytimes.com/2016/06/28/opinion/marine-le-pen-after-brexit-the-peoples-spring-is-inevitable.html

intentions such that one can speak of what it does (not) 'want for itself', enduring in its unity such that one can speak of its 'destiny', and given definition by its demarcation against 'experts'. But other aspects worth highlighting are the following. One sees a clear emphasis on the political over the economic. Brexit is 'above all, a political issue', however much it may be coloured by the debates of 'financial experts'. Budgets are things one should 'determine'. The speaker is making a statement of the primacy of politics.[33] Politics deals with the 'simple and fundamental' things—integers, one might say, rather than the fluctuating fractions of economics (the 'exchange rate'). Echoing the language of the Brexit referendum, British voters are understood to be motivated by an intention to 'regain control', in so doing exercising a 'free choice'. What they seek to escape is something that binds their capacity to make choices—a 'cage', a 'prison', made up of an undifferentiated array of 'commissions and councils'. Like budgets, borders are said to be things that should be shaped by popular will: in the current EU configuration they are symbols instead of powerlessness.

Such themes would become recurrent in the discourse of the Front National (later 'National Rally'), and more generally of parties across Europe challenging those of government and opposition. Mixed in with the aspects typically characterized as populist have been denunciations of the determining influence of economics, the EU as an emblem of *Diktat*, and the need to regain 'control'.[34] The *Alternative für Deutschland*, whose very name of course is a repudiation of the *Alternativlosigkeit* embodied by the Grand Coalition, was founded as an effort to challenge the discourses of economic necessity invoked in Eurozone crisis management.[35] Its breakthrough in the context of migration-related emergency politics in 2015 has often been interpreted as showing the degree to which its support is motivated by identity concerns. As in the extract above, however, it is worth noting that migration tends to be presented as a symbol of the larger phenomenon of powerlessness. 'Uncontrolled migration' is the problem diagnosed, with the suggestion that here is one more context in which control needs to be asserted. The common dominator is the promise of agency.

[33] Cf. Marta Lorimer, *Europe from the Far Right: Europe in the Ideology of the Front National and Movimento Sociale Italiano / Alleanza Nazionale (1978-2017)*, PhD thesis (London: LSE, in preparation).

[34] The head of the Italian League Matteo Salvini evoked the EU in similar terms as a 'straightjacket' and 'cage' (24 June 2016: https://www.italiaoggi.it/news/brexit-salvini-grazie-uk-ora-tocca-a-noi-201606240905281793; 23 January 2018: http://www.repubblica.it/politica/2018/01/23/news/salvini_berlusconi_parametri_ue-187093449/); Italy needs to 'regain control of its borders, its currency, its banks, its agriculture, its commerce, its fisheries...' (25 June 2016: http://www.ilgiornale.it/news/salvini-contro-lue-dei-massoni-litalia-torni-controllare-1275956.html; also 23 January 2018, above). Likewise Farage on Brexit: 'It represents a once-in-a-lifetime opportunity for the UK to throw off the shackles of EU membership and stand proud again in the world' (6 June 2016: https://www.telegraph.co.uk/news/newstopics/eureferendum/12085693/Leaving-the-EU-is-more-important-than-party-politics.html). See also Geert Wilders (Dutch Party for Freedom) on the need to escape the 'EU prison' (26 October 2016: http://www.breitbart.com/london/2016/10/26/geert-wilders-breitbart-britain-brexit-pioneer-others-will-follow/).

[35] http://www.taz.de/!5346303/

Though they may invoke the value of democracy, certainly not all such figures advocate mass involvement in decision-making. Clearly this is one area in which Left–Right distinctions may be relevant. If we restrict the discussion for now to parties of the Right,[36] it seems accurate to say that most are not participatory democrats, and their supporters need not be looking for hands-on control themselves. The appeal to democracy signifies something more basic: the call for a politics guided by volition rather than the dictates of circumstance and scientific reason. Further features typically ascribed to populism, such as an impatience with procedures, are consistent with this emphasis, as things which may obstruct agency or retard it.

Not only are these not always calls for participation, but it is not always the will of 'the people' that is invoked. In a number of these mobilizations one sees a politics of volition not quite graspable in such terms. Think of the irreverent persona of someone like Boris Johnson (Donald Trump too, though he is not part of our story). Though sometimes talking the language of 'the people', such figures do not merely present themselves as responsive to popular will. They present themselves as acting as much on their *own* will, and invite their audience to identify with them for exactly that. In contrast to those elusive officials who rationalize their decisions by things external—socio-economic forces, the demands of peers—the Johnsons of this world imply, by their words and a more general non-conformism, 'I make my own decisions: no-one behind me pulls the strings.' Though very much a repudiation of necessity and an assertion of agency, it is not quite an assertion of *popular* agency – voluntarism would be the better description. Here again one sees how populism as an analytical frame does not quite capture the whole of it.

The promise of agency articulated by such figures may of course be quite insincere. Just as the technocrat may disguise the real discretion they exercise, the 'populist' insurgent may talk the language of change while consolidating the worst aspects of the status quo. They may pose little threat to the economic structures that feed a desire for agency. Moreover, those warming to such messages need not believe that those uttering them will do much to make their lives better.[37] The substantive goals may be vaguely defined, and even when clear they may be objectionable to parts of the very constituency of support they elicit. (Giving support to a programme that goes against one's interests is, after all, one way to reclaim a sense of agency, as it depends on a conscious resetting of

[36] Leftist parties sometimes classed as populist, such as Syriza, are examined in the next chapter.

[37] An illuminating empirical analysis is: Lisa Mckenzie, 'The Class Politics of Prejudice: Brexit and the Land of No-hope and Glory', *British Journal of Sociology* 68 (S1) (2017), esp. pp. 275ff. Analysing the 2016 Brexit referendum she writes (p. 278): 'The people I spoke to did not vote Leave because they thought it would improve their lives. They did so . . . because they just couldn't stand it being the same. Despite the continued arguments around the United Kingdom leaving the European Union that will continue for many more years, as "Peter" from the East End predicated it is unlikely there will be any obvious positive effects for him or most of Britain's disenfranchised urban and regional poor.'

priorities.) Those for whom the fatalism runs deep may doubt whether even such counter-establishment agents can achieve much, and may feel comfortable supporting them precisely for that reason. But credible to their audience or not, such invitations to 'take back control' stand as forthright statements of intent. The point they convey is that here, finally, someone is willing to take a stand against a world where so much is said to be inevitable, and to do what they *want* to do.

The transnational executive brought to the fore by emergency rule presents its critics not just with a political class to denounce, but those we might call the thieves of agency. They are easily portrayed as such by the complexity of their institutional relations, their willingness to reject the language of volition, and their claim to an expertise that will inevitably escape the grasp of most. Consider the impenetrable language in which transnational emergency rule is conducted. As evident in the politics of the Euro crisis, it is replete with technical terms and terminological innovations (e.g. the 'macroeconomic imbalance procedure'). Although heightened by the economic nature of the challenges faced and the prominence of economists in handling them, this may be intrinsic to the mode of rule in question, driven by the concern to address specialized audiences rather than mass publics alone (indeed, to demobilize the latter), and by the need for a language of communication by which to coordinate the activities of dispersed executives. The effect is to construct a field of activity that appears impervious to outside efforts to follow or influence it.

Moreover, the very opaqueness of this activity may invite the suspicion that those involved have more discretion than they are letting on, that there is something even conspiratorial at work. The conditions become ripe for a counter-politics that defines itself by authenticity and undisguised intentions, its willingness to unmask the secrets of power, and by its pursuit of simplification so that agency can be exercised once more.

Not only by inviting the promise of agency does transnational emergency rule foster the outlook associated with populism. A sub-theme to note concerns the temporality of emergency rule. It is part of the exceptionalist way of thinking that unconventional actions are deemed not only necessary but needing urgent execution. Pursued in quick succession, and escaping the rhythms of democratic politics, such measures contribute to a fast-changing political environment. They foster a world of interruption and flux—of sudden policy shifts and institutional shape-shifting. They are conducive to a kind of ontological insecurity of the political realm.[38] It seems reasonable to suppose such conditions suit those who appeal by contrast to points of supposed continuity: to a 'people' or 'nation' said to be defined by its cross-temporal unity.

It is not that this counter-politics defines itself by an orientation to normality and calm. Scholars of populism observe that here too one sees efforts to capitalize

[38] Anthony Giddens, *The Consequences of Modernity* (Stanford: Stanford University Press, 1990).

on ideas of crisis. An atmosphere of emergency may seem advantageous to right-wing mobilizers, not so much to shroud new initiatives with an air of necessity as to up the stakes and make an existing radical agenda seem plausible. The imminent threat, the urgent need to act, an expectation of immediate effects, and an impatience with the procedural constraints that might retard them, are very much themes of these discourses.[39] Consider how Matteo Salvini, head of the Italian *Lega*, characterized the European Parliamentary elections of May 2019: not as mere elections but 'a referendum between life and death, between past and future'.[40] Insofar as populism is a critique of mediation,[41] it is partly the critique of a certain temporality, of a politics equated with hesitation and a lack of urgency. Extreme circumstances, suddenly arising and in need of swift response, are part of the populist imaginary too.

But the narrative role of rupture and dynamic change in right-wing populism is to act as a point of contrast to a trans-temporal collective unchanging in its essential features and ready to take charge of its destiny. It serves to give definition to the articulation of peoplehood and the capacity for agency attributed. Never is the indivisibility of the community and the unity of its will more easily imagined than against the background of rapid political changes that threaten to dissolve it. Michael Freeden identifies the populist trope of a 'self-assumed past time-line that needs to be reinstituted in the face of its rupture'. The claim to sovereignty is partly a claim to '*temporal* sovereignty'.[42] The populist suggestion is that, in a fast-changing world, there remains one last source of ontological security, one point of continuity between past and present. By evoking a collective ready to reassert itself in the face of wider tendencies towards fragmentation, these champions of 'the people' position themselves as the true voices of agency.

In such ways these political currents define themselves in opposition to some of the tendencies most visible in transnational emergency rule. In the prevalence of crisis talk and a certain indifference to existing norms and procedures, the two display a symmetry of sorts. Both are at home in a world of emergencies and urgent threats. But whereas the agents of established executive power exercise their discretion in the name of necessity, their critics from the margins assert that what they propose to do they will do by volition. Charismatic exceptionalism proudly announces the unbound style of rule that technocratic exceptionalism quietly concedes.

Clearly, transnational emergency politics is only one context of the currents in question. There are other conditions too that encourage the desire for agency,

[39] On the place of crisis in populist thought: Moffitt (*Global Rise*, ch. 7); on the theme of the existential threat in far-Right thought: Lorimer (*Europe from the Far Right*, ch. 5).

[40] https://rep.repubblica.it/pwa/generale/2019/05/12/news/europee_salvini_alza_la_posta_un_referendum_su_di_noi_-226120594/?refresh_ce

[41] Bickerton and Invernizzi Accetti ('Populism and Technocracy').

[42] Michael Freeden, 'After the Brexit Referendum: Revisiting Populism as an Ideology', *Journal of Political Ideologies* 22 (1) (2017), p. 4.

including the long-term hollowing of party politics. Emergency rule leads to the intensification of these, but they can exist in its absence. Nor do I want to suggest that the promise of agency is the *only* source of appeal that so-called populists can draw on: there are surely those for whom xenophobic discourses alone are quite sufficient means of enticement (though as noted, resentment, e.g. of immigration, may often be hard to separate from a more general resentment of 'out-of-control' forces). Finally, I certainly do not want to suggest that right-wing politics is the only form in which a desire for agency can be expressed. But accepting these qualifications, it is this aspect—the promise of agency—that provides the clearest and perhaps least appreciated link between the politics of emergency and the responses it typically meets.

Populism as the Emergency-to-Come

A standard observation of research on political authoritarianism is that the effort to hold executive agents to account leads them towards ever more coercive forms of rule. It is not difficult to see how critical responses to the transnational politics of emergency can result in measures intended further to constrain dissent, creating a potentially open-ended cycle. Even where their electoral success is limited, the *possibility* of a breakthrough by challenger parties becomes reason enough for the existing holders of executive power to highlight the threat and to pre-empt it. The spectre of populism becomes the next emergency—a threat with something of the irrational force about it, demanding urgent moves to forestall it. Emergency rule and 'populism' develop in symbiosis.

The legacy of European emergency politics bears out this pattern. One sees measures taken explicitly as efforts to interrupt the rise of populism, including the securitization of dissenting perspectives and the consolidation of rules intended to limit the scope for political alternatives.[43] 'Anti-populism'[44] becomes increasingly central to the discourse and actions of executives engaged in designing the post-

[43] See e.g. Romanian PM Emil Boc, July 2011 (https://eurocrisislaw.eui.eu/wp-content/uploads/sites/44/2019/05/Romania.pdf), using the spectre of populism to justify inserting constraints on economic policy into the country's constitution: 'I will tackle the provision of Article 138-1 regarding the state financial policy, which requires, for the first time in the Romanian Constitution, a limit of the deficit at 3% and of the public debt exceeding 60% of GDP. Such a constitutional provision is not new, or completely new, to European constitutions. We already have the precedent of Germany in 2009, which included a provision almost exactly as ours on the budget balance and the discussions are well advanced in France, for the purposes of such an approach to the provision in the Constitution of certain rules to keep the financial balance. I think that such a rule is absolutely necessary to be imposed at the constitutional level. I say this especially in view of the lessons the crisis have taught us about how vulnerable governments and countries are facing populism, demagogy and the policies responsible for buying votes with public money.'

[44] On 'anti-populism': Yannis Stavrakakis, Giorgos Katsambekis, Alexandros Kioupkiolis, Nikos Nikisianis, and Thomas Siomos, 'Populism, Anti-populism and Crisis', *Contemporary Political Theory* 17 (1) (2018), pp. 4–27.

crisis order, with the term populism applied increasingly indiscriminately. The structure of the transnational realm adds some distinctive dynamics, including how experiences in one country can be invoked as lessons in another. Even if the populist challenge is often national in scope, it is easily transnationalized by those seeking to play on the fear of it.

But our preceding account allows us better to understand what is going on. What makes the spectre of populism so problematic for established authorities is, on the one hand, that it interferes with efforts to draw a line under emergency rule and to consolidate the institutional and policy regimes intended to stand as the new normal. By advancing the promise of agency, such agents reintroduce uncertainty into the political sphere, and open the possibility that changes made under the sign of emergency may be undone.

On the other hand, and perhaps more disruptively still, the appearance of such groups threatens to undermine the very logic of necessity by which these changes were first introduced. Emergency rule relies on presenting the extraordinary measures undertaken as the only plausible way to stave off the breakdown of order. Its strategy of securitization is to emphasize the threats it contends with as profound ones, and to present alternative responses as destined to maintain or amplify the crisis at hand. Yet challenger parties making the promise of agency raise a challenging prospect: not only that alternative policies might be pursued in future, but that they might be introduced *without* reigniting the conditions of crisis used to justify yesterday's exceptional measures.

In the economic context, one thinks here of departures from the austerity model—departures which insurgent politicians have certainly hinted at, credibly or not. Were such promises to be followed up without largescale disarray—i.e. with economic activity sustained and public order maintained—then they would cast doubt on the necessity of earlier crisis measures. In this way the central claim of executives engaged in emergency rule to be doing what is required would be left exposed. The suspicion could then arise that these exceptional measures were a choice rather than a last resort. The authority of those who instituted them, and in particular their persuasive capacity to press for such measures in future, would accordingly be severely impaired.

The challenge presented by 'populist' parties is thus two-fold. In the forward-looking direction, it raises the possibility of policy changes unpalatable to established executives. In the backward-looking direction, it questions the emergency rationale invoked for measures previously instituted, and thus also questions the authority of those who undertook them—those claiming particular insight based on crisis expertise. It is the *promise of agency* associated with the currents of populism that is the heart of the challenge: the promise not only to do things differently in future, but to expose the agency already wielded in the past.

It is unsurprising then that the tendency in EU policy circles since the early 2010s has been to redouble efforts to minimize the possibilities available for the exercise of agency. Though scope for discretion is retained for the more powerful, it is cloaked behind mechanisms that disguise and anonymize it.[45] Most clearly in the economic rules associated with the European Semester, the apparent aim is to create unchanging rules and coercive mechanisms that more or less apply themselves, and that create impediments for those who might chart a different path. In a polity composed of multiple electoral systems and thus multiple opportunities for the promise of agency to be voiced, the effort to restrain it is all the sharper.

Quite aside from their substantive effects, such moves reaffirm an image of the EU as self-referential, as a set of mechanisms that carry on regardless. The sense of agency denied, and thus the incentive for figures to emerge claiming the capacity to restore it, is reinforced. The symbiosis of emergency rule and the anti-establishment response it provokes rotates, I suggest, on the question of agency.

One implication of the argument is that the politics of 'populism' is propelled at least as much by the weaknesses of representative democracy under conditions of emergency rule as by any popular attraction to the specific features one tends to associate with it (non-pluralism, anti-elitism, etc.). In the terms of the debate, it is a problem of supply rather than demand. The appearance of challengers is a symptom of wider failings of institutions and political culture, intensified in the form of the transnational politics of emergency. In a political setting where the scope for agency is rejected and constrained, 'populists' happen to be those taking a stand, rhetorically at least, for a politics of volition rather than necessity. That they should draw their support on this basis is consistent with one of the recurrent findings of empirical research on support for populist parties—that it cuts across class lines, extending to both middle and working classes.[46] Given the shared experience of democratic evisceration, the promise to restore an agency denied is likely to find broad appeal. There will be a sizeable constituency for it, well beyond the circles of xenophobia, racism and anti-politics, for as long as governing in major policy areas proceeds by the principle of necessity.

Another implication of the argument however is that one may conceive responses to emergency rule which carry public resonance yet which are *not* coupled with the more problematic features of right-wing populism. Too often populism is analysed as part of an exclusive dichotomy—e.g. as 'responsive' politics, contrasted to 'responsible'. Such a perspective leads to the conflation of diverse forms of dissent against established power into a single cluster of opinion, a move tarring some by association with others and embedding the notion that serious opinion comes in only one form. But things are different if we see a significant driver of these movements as lying in their suggestion of political empowerment. For it is by no means only in this form that the promise of agency can be articulated: the connection between the two is circumstantial.

[45] Menendez ('Crisis of Law').

[46] Mudde and Rovira Kaltwasser (*Populism*).

The question becomes whether the interplay of emergency rule and exclusionary populism may be interrupted by political forces reducible to neither. Such a third form would be one that carried over the promise of agency expressed in populism, challenging thereby the depoliticizing tendencies of emergency rule, while distinguishing itself from the more sectarian and exclusionary features associated with right-wing populism. This possibility is explored in the following chapter.

Conclusion

Persistently engaged in the politics of emergency, the EU has come to rationalize some of its most important decisions as urgent measures to handle a crisis. Whether one sees these moves as mainly opportunistic or dictated in some measure by circumstance, they amount to a governing formula based on the principle of necessity. With the suggestion that decisions are compelled rather than chosen, advancing measures of last resort, this formula consolidates an image of the EU as the negation of political agency. It is an extension of, but also a distinctive development upon, long-standing tendencies towards the denial of agency, both amongst parties of government and the officials of technocratic institutions.

'What a Prince or Republic does of Necessity, should seem to be done by Choice.'[47] So advised the realist Machiavelli, on the grounds that only those rulers who claimed responsibility for their actions would be able to court popular support. Allegiance needs agency to attach to. The pitfalls of arguments from last resort would seem to be well understood by a number of insurgent movements of the present period. I have indicated how the practices of emergency rule invite their mirror opposite: a politics defined by its rejection of necessity and the celebration of volition. The mobilizations commonly studied under the heading of 'populism' bear this feature prominently, and it is plausible to suppose a good portion of their wider appeal comes from the promise to restore political agency—a promise independent of the other features typically ascribed to them, and which the term populism can serve to obscure. The distinguishing feature of many such movements is their insistence that politics should be about more than adaptation to socio-economic demands, that it should set its own priorities rather than merely react. This is a repudiation of political acquiescence. To dwell on parallel characteristics distinctive to populism may in some ways be falsely reassuring, since it overlooks how an equally enticing promise of agency may be offered by considerably more dangerous groups, fascists amongst them.

[47] This is the title of Machiavelli's *Discourses* LI.

But it also overlooks that there is positive potential here too. If the disavowal of agency is one of the enduring failings of the European political space, its constructive resolution will depend on the emergence of those willing to propose consequential interventions in the socio-economic sphere. Amongst those who currently present themselves as figures of agency, this willingness may often be superficial. They may rail against socio-economic necessity without advancing policies that might genuinely test it. Much like the crisis managers they oppose, their interventions may be targeted at the expressions of malaise rather than its causes. But even if their prescriptions are misdirected, it seems important to avoid casually ascribing what support they muster to their least desirable features. Anti-elitism, cultural othering, and a certain kind of anti-pluralism are features that can support only a regressive political cause. The promise of agency, by contrast, is an idea that can be detached from them and embraced by others.

7
Principled Disobedience

A theme of the book has been that improvisations in the name of crisis management may be coupled with efforts to lock in some of the measures adopted. Warding off further disorder may be said to demand heightened constraints on agency. Though some may be spared these controls—the design of the rules leaves more powerful agents undisturbed—for those in a subordinate position they may have considerable binding force. In the contemporary EU, governments and parliaments have been amongst those increasingly hemmed in. An effort to normalize the order arising from the EU's brush with emergency rule has taken the form of limiting these institutions' room for manoeuvre.

An important implication is that those wishing to reshape the order arising may be compelled to do so by rule-breaking means. As policies are embedded, challenging them by procedural means becomes harder. This matters not least because an order shaped under the sign of emergency may be highly imperfect. Built in haste by agents whose mandate for far-reaching change is slim, its rules may be quite problematic. But if one rightly has concerns about the unbound character of emergency rule, what is one to make of the kinds of transgression that may be required to overturn its legacy?

The politics of disobedience is the subject of this chapter. I examine the possibilities for disobedience in a transnational order, in particular as pursued through institutions, and the agents best placed to take advantage of them. Party-led forms of disobedience are given particular attention. Also in focus are the normative questions raised. Drawing on criteria inspired by theories of civil disobedience, I discuss the basis on which a certain kind of disobedience can be distinguished from the measures it is intended to challenge, as well as from other destructive forms of exceptionalism. I develop an account of what 'principled' disobedience in the transnational setting might entail.

The Syriza movement that came to power in Greece in January 2015 was one formation in whose actions one saw something of this stance. Prior to entering government, it had articulated a critique of (para-)EU policy-making as a critique of emergency rule. As the party's leader Alexis Tsipras put it in early 2014: by using the crisis as a mere pretext, the neoliberal establishment of Europe is attempting a violent re-arrangement of Europe's division of labour... [We are planning] to put an end to the Caesarean attitude of the European Union, as Gramsci would have put it, to domination as a tool for managing the crisis.... The neoliberal establishment in Europe has managed this crisis, not in order to resolve

Politics of Last Resort: Governing by Emergency in the European Union. Jonathan White, Oxford University Press (2020). © Jonathan White.
DOI: 10.1093/oso/9780198791720.001.0001

it, but in order to revise the European postwar political economy itself. In order to instigate an assault by capital against labour, in order to dismantle both the rights and the hard-earned gains on the part of the forces of labour throughout Europe's continent.... For that purpose it has succeeded in overriding the very institutional framework of the European Union. It has strangulated democracy. It has disregarded the multifariousness of institutions at a national level. It has gagged the peoples, forcing upon them the straitjacket of austerity, discipline and deregulation.[1]

Like the responses discussed in the previous chapter, such words express impatience with a politics that refuses meaningful choices in the name of crisis management. Here was a party expressly defining itself by the promise of agency in opposition to the straitjacketing of democratic processes, and intent on exposing the discretion concealed behind discourses of necessity. Yet rather than addressed to the interests of an ethnos on the model of right-wing populism, here was a party articulating the grievances of the *peoples* of Europe and an assault on rights across the continent. Rather than positioning itself as the enemy of procedures and fetishizing a stance of pure volition, here was a party seeking to defend the EU institutional framework from assault. As we shall further examine, here was a generalizable critique of the transnational politics of emergency and a commitment to overcoming the deficiencies that made it possible, coupled with the germ of a strategy of disobedience for tackling it.

In light of how the Syriza experiment played out in 2015, when the threat of disobedience was evoked only to be soon discarded, the prospects for an intervention of this kind in the EU may currently seem remote. Indeed, given the scale of the power resources that can apparently be applied to forestall it, further consideration may seem wholly premature. Yet this has not stopped central figures in the Syriza experiment continuing to issue calls for a politics of 'civil, civic and governmental disobedience' in the EU, the theoretical basis of which deserves further consideration.[2] Here is a context in which to trace out the suggestion, made amongst others by Honig, that emergency politics can generate opportunities for democratic engagement, in amongst the challenges to democracy it poses.[3] Moreover, what the Greek episode also showed was the confidence with which charges of irresponsibility are levelled at those who toy with the prospect of defaulting on their obligations, and the symbolic force of appeals to 'play by the rules'. It seems appropriate to consider whether there are not circumstances in which 'irresponsibility' may have its place, and to investigate

[1] http://www.alexistsipras.eu/index.php/10-speeches/49-alexis-tsipras-speech-rome-7-02-2014?showall=&start=1

[2] Yanis Varoufakis, 'Europe's Left after Brexit', 8 September 2016, https://yanisvaroufakis.eu/2016/09/08/europes-left-after-brexit/; also DiEM25's July 2017 'Guide to Constructive Disobedience': https://diem25.org/constructive-disobedience/

[3] Honig (*Emergency Politics*), p. xvi.

the criteria that might regulate it. The final section reflects on what these remarks imply for the larger legitimacy of emergency rule.

With Far-Right political forces in Europe never far from electoral advance, these questions have added significance. Dissent towards rules made in Brussels is a cornerstone of these parties' appeal, and there is no reluctance to invoke defiance of EU rules, either as rhetoric or strategy. Disobedience, verging on sabotage, is a game the Far Right can play, in the service of an undifferentiated hostility to transnational institutions. Taking critical distance from these positions, and challenging their monopoly of opposition to the post-crisis EU, requires that we reflect more closely on the criteria that separate legitimate forms of disobedience from the dubious.

Why Disobedience?

The political order emerging from Europe's emergency politics of the 2010s included a significant body of new binding demands. One need only recall the stipulations included in the Fiscal Compact concerning how states are to balance their accounts. The requirement that Eurozone members give constitutional status to a 'debt brake' on public finances has drawn particular attention, as has the requirement that states submit their budgets to the Commission for approval.[4] Such demands confront states of the Eurozone, some more than others, as significant constraints, as the standoff between Greece and the Council in summer 2015 made clear.

As many have argued, it would be misleading to see these new demands as amounting to the strengthening of the rule of law. There remains significant space for unchecked discretion for those who have the power to claim it, whether they be the governments of the largest member-states or supranational agents such as the Commission and ECB. Furthermore, the sheer variety of measures instituted, and the quite different legal foundations they rely on, compromise the coherence and intelligibility of the resulting framework.[5] As a distinguished legal scholar observes, EU law has been devalued by cynical use, employed as a stabilizing tool without sensitivity for norms and generalizable aims, and in an uneasy relation with public international law. While rules are increasingly abundant, the larger process they give rise to is one of de-legalization.[6]

It is often observed that the unconventional measures of emergency rule tend to become quasi-permanent. The prospects for their roll-back are difficult, particularly if the underlying crisis is widely seen to be long-lasting.[7] One reason is that

[4] See esp. Art 3(2) TSCG. [5] Kilpatrick ('Rule of Law').

[6] See Joerges in Chalmers et al. (*End of the Eurocrats' Dream*), esp. p. 308.

[7] Kreuder-Sonnen ('Beyond Integration Theory').

the measures instituted may come to seem necessary to prevent reopening a crisis: the will to undo them is weak. But where there *is* the will to contest them, by what channels can it find expression?

I am interested in how such rules and conditions may be challenged rather than avoided. Clearly an exit from the EU is an option available to those seeking to escape such demands, more or less feasible in the particular case. But such a response disputes the application of these constraints to particular agents rather than their existence as such. Extrication, not contestation, is the logic of exit. Such moves do little to challenge EU rules in the round, as the Brexit process may suggest. Indeed, by reinforcing the dominant division of opinion on EU matters as that between pro-membership or against, it may foster uncritical support for the Union as the main alternative to exit sentiment.[8] Exit can reinforce what it leaves behind. So what options exist instead for *contestation*?

The first mode by which the accretions of emergency rule may be challenged is by *ordinary legislative means*. In the EU, this conventionally involves an intergovernmental decision by state representatives in the Council, influenced in some measure by the European Parliament, and in response to proposals formulated by the Commission. Such an approach is in principle compatible with legal and political norms: it need not mark a renewal of the exceptional means associated with emergency rule, nor need it feature disobedience of the kind discussed below. This is the approach advocated by many sympathetic critics of the contemporary EU, who consider its arrangements unsustainable yet who do not wish to advocate its dissolution.[9]

There are good reasons why this mode of contestation is preferred. It aims to tap the legitimacy of established institutions, national and supranational, and to uphold the principle of a rule-bound order while challenging the specifics of the order as found. It reaffirms the EU's long-standing self-image as the achievement of 'integration by law'. A polity that undergoes such a transition promises to emerge from emergency rule with its procedural authority renewed.

Appealing as it may be, such an approach faces sizeable obstacles, particularly in a setting characterized by the diffusion of power. Reaching a decision by this method requires, depending on the policy area, the unanimity of EU member-states or the consensus of a large majority. Such hurdles are not easily crossed. To the extent that states retain the same governments that instituted emergency measures in the first place, a reversal requires their willingness to overturn their own decisions. To the extent national parliaments were involved in approving these measures, even a change of government may be insufficient to change the

[8] Polling data indicating a rise in pro-EU attitudes across Europe during the Brexit process would seem to support this: http://www.europarl.europa.eu/news/en/press-room/20181016IPR16213/brexit-effect-public-opinion-survey-shows-that-eu-is-more-appreciated-than-ever

[9] E.g. Habermas (*Lure of Technocracy*).

dynamics. Without additional pressure of the kind we shall discuss, a constellation for change is unlikely to appear.

Those who see the EU as subject to the hegemony of a single state—Germany—may feel that the direction of EU decision-making hangs mainly on German domestic politics and the composition of government which arises.[10] It should be enough to recall though that EU decision-making in recent years has not always mirrored the preferences of the German government. Especially the interventions of the IMF and ECB have challenged it at key moments, as have the positions of other national governments.[11] The Troika chariot was not easily reined in. The course of EU politics depends on considerably more than the politics of just one country, and this variety of veto players makes the reversal of policy more difficult.

A *judicial* response to emergency rule is the second kind that tends to be invoked. Courts are typically accorded a prominent role by scholars of emergency politics. Rarely may they be in a position to *prevent* emergency measures, but other functions are commonly foreseen. They may draw public attention to threats to legality, acting as 'weathermen' while emergency rule is ongoing.[12] Later they may be in a position to challenge the legality of the measures adopted and apply pressure for their reversal or modification. Where the path to change by ordinary legislative means is blocked, a judicial channel to reform may remain open.

A judicialized order such as the EU would seem to display the appropriate conditions for a court-led challenge to emergency rule. Yet as noted in Chapter 4, the CJEU has generally been unwilling to contest measures introduced for crisis management. Indeed, the main judgements of relevance (the 2012 Pringle case on the European Stability Mechanism, and the 2015 ruling on OMT) served instead to approve those measures. The response of the CJEU in this period was to reinforce the discretion of the powerful rather than challenge it.[13] What are we to make of the acquiescence of the court best placed to contest transnational decisions from a transnational perspective? Is this another case of the 'judicial deference'[14] that national courts so often display towards executive discretion under the sign of emergency and by which they abandon their critical function?

One of the structural problems for a judicial response to emergency rule is that, in the transnational context especially, the executive actions in question are not

[10] On German hegemony: Ulrich Beck, *German Europe* (Cambridge: Polity, 2013). On the European significance of German politics: Habermas (*Lure of Technocracy*).

[11] David Schäfer, 'A Banking Union of Ideas? The Impact of Ordoliberalism and the Vicious Circle on the EU Banking Union', *Journal of Common Market Studies* 54 (4) (2016), pp. 961–80; Tooze (*Crashed*).

[12] David Dyzenhaus, *The Constitution of Law: Legality in a Time of Emergency* (Cambridge: Cambridge University Press, 2006).

[13] Kilpatrick ('Rule of Law'); Scicluna ('Integration through the Disintegration of Law').

[14] Kim Lane Scheppele, 'The New Judicial Deference', *Boston University Law Review* 92 (2012), pp. 89–170.

easily classed as illegal. The 'incompleteness' of the EU constitutional order means that, however dense the framework of rules in certain areas, there are zones of action over which the binding capacity of law is quite limited. The willingness of executives to step outside the EU legal structure when seeking to institute measures in the face of resistance further limits the reach of the EU judiciary. Intergovernmental compacts are never likely to be effectively challenged by the CJEU. Where EU law poses an unwelcome constraint, the more powerful agents can avoid it.

But this can only be part of the story, given that scholars do indeed observe illegal actions in the making of the new regime for the post-emergency era,[15] and more generally given the past willingness of the CJEU to adopt an activist stance and claim jurisdictional authority in areas where its status is ambiguous.[16] Two further explanations of acquiescence are these. First, there is the general problem that, were a court to conclude that major illegal acts were conducted over a sustained period, it would put in question the credibility of the legal order of which it is guardian. Especially for an international court, inclined to bolster what remains a contested legal order, this would be a major concession. Tempting to conclude instead that the initiatives taken were consistent with law.

Second, there is a problem of epistemic authority. Where emergency measures rest on claims to expertise, the contentious agent sets itself against this expertise. Such a stance is a difficult one to assume for the small number of persons charged with arriving at a judicial decision. Few judges can be expected to challenge such measures when the consensus of a much larger group is that to do so would reignite the crisis. Especially when the relevant expertise goes beyond conventional scientific knowledge to a more intuitive form of *know-how*, an attitude of trust is likely to be preferred. The deference accorded to the 'expertise' of the ECB was one of the notable features of the CJEU's legal reasoning of the period.[17] As one observer summarized in relation to the *Pringle* decision, 'was there ever any doubt that the Court would rule as it did given the stakes involved?'[18]

This is not to say courts are unimportant in the contestation of crisis measures. Some national courts have been more confrontational than the CJEU, albeit from a national perspective.[19] But one may doubt that a judicial response alone is sufficient. In any case, once a court finds actions to be legal, the decision is unlikely to be reversed: the window of opportunity is brief.

[15] E.g. Fischer-Lescano (*Human Rights*); Joerges ('What is Left?').

[16] Kilpatrick ('Rule of Law'), p. 352.

[17] See esp. the OMT decision of 16 June 2015, and before it the Opinion of the Court's Advocate General, delivered 14 January 2015.

[18] Nicole Scicluna, 'Politicization without Democratization: How the Eurozone Crisis Is Transforming EU Law and Politics', *International Journal of Constitutional Law* 12 (3) (2014), p. 564.

[19] See e.g. the Portuguese and German national courts; see also Joerges in Chalmers et al. (*End of the Eurocrats' Dream*).

Besides an ordinary legislative approach and a judicial one, there is one further response that shares with these the characteristic of not requiring further rule-breaking. We might call it *malicious compliance*. Its point of departure is that the assemblage of rules and commands arising from emergency rule are likely to display tensions, both with each other and with the pre-crisis order. The speedy and ad hoc fashion in which they are introduced generally undermines the formal coherence of the resulting framework, while the tailoring of measures to very specific concerns means they may rapidly become obsolete as the context changes. This may suggest that what is needed to overturn them is no more than dogmatic adherence to them so that their internal tensions undermine them.

Malicious compliance is a strategy usually associated with trade unions contesting workplace conditions. Such tactics as *work-to-rule* involve adhering to the letter of contractual obligations so as to reveal their dysfunctionality and/or dependence on discretionary interventions. The approach is in principle transferable to a larger political unit. In the EU context, the demands associated with the Fiscal Compact are a potential target, not least since the austerity measures they entail are widely observed to be unsustainable.[20] Even the broadly sympathetic magazine *The Economist* has had reason to note that strict adhesion to Eurozone banking rules could have fundamentally destabilizing consequences. With admirable prudence, if a want of imagination, it concludes: 'there is no point in following rules to the letter, if doing so leads to the demise of the single currency'.[21]

As a form of contestation, one distinctive plus of malicious compliance is its invulnerability to sanction. Because it aims to challenge rules from within, it can be pursued without compromising its protagonists before the law, and without giving executives further reason to introduce constraining measures. Furthermore, because malicious compliance is essentially reactive, it requires no programmatic agreement amongst those practising it about the alternatives they would like to see. It can be pursued by actors of contrasting political orientation.

Malicious compliance is an under-theorized notion that invites further scrutiny, but a significant shortcoming should be noted: the tensions it aims to highlight can typically be alleviated or resolved in multiple ways, only some of them socially desirable. In principle they can be addressed by sacrificing good rules and norms just as well as bad ones. As a stand-alone strategy, malicious compliance is silent on the question of *how* tensions in the rules framework are to be addressed. The effort to ensure normative and not merely functional criteria are brought to bear on the resolution of such tensions is likely to require more focused forms of contestation.

[20] E.g. Blyth (*Austerity*).

[21] See *The Economist* leader, 'The Italian Job', 9 July 2016, p. 9.

Disobedience in a Federation

To contest the order emerging from EU crisis politics in a rule-abiding way is difficult. The implication is that opponents may be pushed to engage in acts of transgression bearing some similarity with the manoeuvres by which these measures were first instituted. Indeed, it may be that acts of contestation entail the breach of rules *more* explicitly than emergency rule itself, since—in the transnational context especially—they must proceed within a rules framework shaped at least partly by the preferences of those powers wielding discretion.

Not only may transgressions be hard to avoid, but on one reading it is intrinsically desirable that at least some measures be challenged in this way. One aim of contestation tends to be to raise critical public attitudes, and open disobedience may be suited to this.[22] If emergency rule relies on securitization, such that exceptional measures are advocated as necessary responses to urgent threats, then disobedience can be seen as a step towards *de*-securitization and re-politicization.[23] If it can be shown that breaching such rules does not produce the immediate chaos that authorities warn of, opposition agents thereby challenge the logic of emergency that formed their rationale. The breach of rules thus has a performative function: it contests the necessity of those rules, and thereby contests the symbolic power used to establish them. As observed in the previous chapter, this is one of the ways many so-called populist parties pose a challenge to the exceptionalism of established executive powers—but it is a challenge that derives from the promise of agency they make, not the exclusionary features of right-wing populism itself.

Disobedience is also about positively shaping the attitudes of citizens towards peers. The views people hold about the inclinations of others is an important influence on how far they seek common cause or instead tend to individualize the predicaments they face.[24] When executives successfully invoke the principle of necessity to reshape the political landscape, not only do they convey the impression that alternatives are unfeasible. By rendering contestation less visible, they give citizens reason to doubt the critical attitudes of their peers, both in their own state and abroad. They foster the conditions for a 'spiral of silence', whereby those with dissenting views become reticent because they think others do not share their outlook.[25] Disobedience is about breaking the spiral of silence. Through a public demonstration of commitment, its practitioners make visible the enduring reality

[22] Cf. William Scheuerman, 'Recent Theories of Civil Disobedience: An Anti-Legal Turn?', *Journal of Political Philosophy* 23 (4) (2015), p. 5.

[23] On de-securitization: Claudia Aradau, 'Security and the Democratic Scene: Desecuritization and Emancipation', *Journal of International Relations and Development* 7 (2004), pp. 388–413.

[24] Jonathan White, 'The Politics of Other Citizens', *Citizenship Studies* 14 (4) (2010), pp. 411–27.

[25] Elizabeth Noelle-Neumann, *The Spiral of Silence: Public Opinion—Our Social Skin* (Chicago: University of Chicago Press, 1984).

of dissenting views. The statement of commitment is all the more credible for the fact that it extends to a willingness to break rules: here again, the fact of transgression has an intrinsic significance, beyond whatever changes it leads to.

Is disobedience a relevant notion in a voluntary association such as the EU? Members of a transnational polity are said to be there by their own consent—does the exit option remove the grounds for disobedience? There are normative aspects to this question we shall return to, but in a practical sense the exit option does not fundamentally change things. Unless there are procedures available for expelling the disobedient (which exist neither in the EU in general nor the Eurozone specifically), disobedience remains a viable option to the extent that those seeking change on balance prefer membership over exit.

Discussions of the politically motivated transgression of rules tend to focus on street-level direct action. Civil disobedience is typically construed as occurring outside the institutions of representation, displaying for this reason virtues of spontaneity and horizontality.[26] This concept of disobedience is not without relevance to the transnational sphere, given the mobilizations associated with the *Indignados, Aganaktismenoi,* Occupy and their successors. Efforts to coordinate such activities across Europe as part of a common agenda of civil disobedience have been pursued by those working under the heading of a 'Plan B for Europe'.[27]

But the transnational context affords a less discussed form of disobedience, one waged *within* institutions. To the extent that power is diffused rather than concentrated at an agreed centre, the decisions enacted through some institutions can be contested by agents housed in others. To hold a position of government at the national level is to hold a position from which to oppose decisions made transnationally. The agents of disobedience may therefore include those located in executive and legislative institutions.[28]

[26] Celikates offers an applicable definition of civil disobedience: 'an intentionally unlawful and principled collective act of protest (in contrast to both legal protest and "ordinary" criminal offenses or "unmotivated" rioting), with which citizens—in the broad sense that goes beyond those recognized as citizens by a particular state—pursue the political aim of changing specific laws, policies or institutions' (Robin Celikates, 'Rethinking Civil Disobedience as a Practice of Contestation—Beyond the Liberal Paradigm', *Constellations* 23 (1) (2016), p. 39). Cf. William Scheuerman, *Key Concepts: Civil Disobedience* (Cambridge: Polity, 2018).

[27] https://www.euro-planb.eu/?page_id=96&lang=en

[28] While ideas of civil disobedience have rarely been applied to the EU institutions, on disobedience by *judiciaries* see Julio Baquerro-Cruz, 'Legal Pluralism and Institutional Disobedience in the European Union', in Matej Avbelj and Jan Komárek (eds), *Constitutional Pluralism in the European Union and Beyond* (Oxford: Hart, 2012); Mathias Kumm, 'The Moral Point of Constitutional Pluralism: Defining the Domain of Legitimate Institutional Civil Disobedience and Conscientious Objection', in Julie Dickson and Pavlos Eleftheriadis (eds), *Philosophical Foundations of EU Law* (Oxford: Oxford University Press, 2012). On state civil disobedience: Antonio Franceschet, 'Theorizing State Civil Disobedience in International Politics', *Journal of International Political Theory* 11 (2) (2015), pp. 239–56; on citizen disobedience and cross-border mobility: Rikke Wagner, 'Transnational Civil Dis/Obedience in the Danish Family Unification Dispute', *European Political Science Review* 7 (1) (2015), pp. 43–62.

This notion of institutional power as a site of disobedience is one way to get a handle on the actions of Syriza in the first half of 2015. One would overlook something essential if one saw the Greek governing party's actions as just state diplomacy in defence of national interests: the ends in play were more far-reaching. One sees this clearly also in the movements later emerging from the failure of Syriza's efforts. The DiEM25 movement, initiated in 2015 by former Syriza figure Varoufakis, proposed in one of its early programmatic statements 'a pan-European movement of civil and governmental disobedience with which to bring on a surge of democratic opposition to the way European elites do business at the local, national and EU levels'.[29] In contrast to EU withdrawal, the suggestion was that the regime emerging from the crisis period could be challenged from within. Advocated was 'a clash with the European establishment based on a campaign of wilfully disobeying the unenforceable EU "rules" at the municipal, regional and national levels while making no move whatsoever to leave the EU'. In keeping with the early policy of Syriza, 'rebel governments and finance ministers' were presented as some of the key agents of disobedience, but the stated aim is to involve governmental institutions at all levels.[30]

As such calls suggest, contesting power through institutions has some qualities unmatched by street-level disobedience. The EU's long-established insulation of decision-making from popular interference means that systemic challenges to it, more than simply highlighting critical opinion, may need to disrupt its functioning too. The demands associated with the European Semester—on public spending, on the release of economic data—are demands that have street-level effects but which in the first instance bind governments rather than citizens, and it is the actions of the former that determine their success. The same holds of certain rules of procedure that support their implementation—e.g. the non-public character of discussions concerning the meeting of budgetary requirements. These can be challenged directly only on an institutional footing.

Disobedience outside the nation-state is not without its challenges, conceptual as well as practical. One of the most significant concerns its *addressee*. If principled rule-breaking involves an effort to convince a larger public of the unsustainability of the status quo, who is that public in this context? Clearly there is a nationalist or statist response to this question: the public is that of the nation-state, incited to see the EU more critically and to call for changes in the state's relation to it. But in a transnational polity there may be strong reasons of both principle and strategy why agents of disobedience should want to construct their constituency as a transnational one. The question would be whether there is a cross-border social formation that can support such claims. Can one expect receptiveness to notions

[29] Varoufakis ('Europe's Left after Brexit').
[30] Varoufakis ('Europe's Left after Brexit'); cf. https://diem25.org/constructive-disobedience/

of a European interest, and enough shared moral intuitions to permit a common critique of the status quo?

A strategy of disobedience must assume that, even if such a constituency is weakly formed, it can be cultivated through the process of contestation itself. The gamble is that a convergence of outlooks can emerge *through* political mobilization rather than act as its necessary precondition, that disobedience can '*create* the European demos that will bring about Europe's democracy'.[31] The plausibility of such views has been much debated and we need not revisit the European 'demos debate' here.[32] But it is clear that disobedient agents in the transnational setting lack some major resources by which to shape a constituency for their claims. For one, the absence of a written constitution articulating shared norms in a popular idiom deprives them of a reference-point by which to highlight the failings of the status quo. Unlike those in a constitutionalized setting, they will struggle to point to codified commitments that the existing order fails to honour. Whether such barriers can be overcome is something that only the practice of disobedience can ascertain. Efforts to decide such questions in advance by sociological means—by reference to divisions of language, reported feelings of identity, and the like—are always liable to overstate the determining effect of existing structures.

Principled disobedience as an *institutional* practice also raises a challenging question concerning the status of its agents. Conceived at a street level, disobedience is the action of individuals exercising their moral judgement and bearing the attendant risks of punishment. If one extends the concept to an institutional setting, the worry will be that one ends up ascribing a dubious moral capacity to institutions.[33] I think one can avoid this, but only by emphasizing the role of intermediary actors such as parties and movements. To the extent that such associations are defined by shared purposes and genuinely express a form of collective intentionality, they are agents to which judgement and responsibility can be ascribed. To the extent that they display deliberation, their collective intentionality is reflexive in character.[34] When such agents gain majority control of institutions, they establish the kind of 'authorship' of decision-making that less cohesive entities with no claim to programmatic unity—including the state itself and its organs of bureaucracy—are themselves little able to offer.[35]

[31] Varoufakis ('Europe's Left after Brexit').

[32] For recent overviews: Fabio Wolkenstein, 'Democracy, Transnational Partisanship and the EU', *Journal of Common Market Studies* 56 (2) (2018), pp. 284–99; Daniel Innerarity (2014), 'Does Europe Need a Demos to Be Truly Demoratic?', *LSE Europe in Question Series (LEQS)*, No. 77.

[33] William Scheuerman, 'Crises and Extra-Legality: From Above and from Below', in Poul F. Kjaer and Niklas Olsen (eds), *Critical Theories of Crisis in Europe: From Weimar to the Euro* (Rowman & Littlefield, 2016), p. 240.

[34] On intra-party deliberation: Fabio Wolkenstein, 'A Deliberative Model of Intra-Party Democracy', *Journal of Political Philosophy* 24 (3) (2016), pp. 297–320.

[35] As should be clear, this argument is distinct from one concerning 'state civil disobedience': I refer to disobedience 'through' institutions but not 'by' them.

Many ruling parties today are rather less than this: they are the 'hollowed' vehicles of a political class, complicit in the practices of governance and emergency rule.[36] Our remarks can only apply to parties organized on a more collective basis. Note though that this hollowing of parties is in some measure the effect of diminished expectations of what partisanship can achieve under contemporary institutional conditions. Where members take seriously the possibility of principled disobedience through institutional channels, and thereby open a new front for partisan agency, they gain renewed reason to develop themselves as cohesive agents of the kind described.

A lingering suspicion may be that disobedience, especially when drawing on the capacities of existing institutions, will be conservative. It can seek merely to correct some obvious policy failings while leaving deeper structures untouched. While the kind we are interested in aims first to reverse the most egregious legacies of emergency rule, this does not mean it cannot be tied to a wider agenda of change. In contrast to a Rawlsian focus on correcting the specific failings of a 'nearly just' order,[37] the stated objective of DiEM25 for instance is the EU's redesign, and, should this fail, its 'collapse or disintegration' in a way that does the least political damage.[38] Although existing institutions may be part of the means, the goal is to alert a wider public to deep-lying failures in the existing order. Disobedience can thus be the leading edge of a programme that aims ultimately at transformation.

Evidently the risk in entertaining ideas of disobedience through institutions lies in the licence they give for exceptionalism to those already enjoying considerable powers. Though we may be far from a world in which office-holders carry a monopoly on the legitimate use of violence, clearly they wield a coercive capacity that can be put to sectarian ends. One needs to consider more closely the basis on which disobedience can be circumscribed.

Normative Foundations of a Disobedient Politics

When might efforts to challenge the order arising from emergency rule, with the infringement of rules and commands this may entail, be considered acts of principle? In what sense is there scope for such moves to be viewed differently from the unconventional measures they oppose, as well as from the anti-establishment orientations of the populist Far Right? Contemporary discussions of the normativity of civil disobedience, adapted to the transnational context, suggest

[36] Peter Mair, *Ruling the Void: The Hollowing of Western Democracy* (London: Verso, 2013).

[37] John Rawls, *A Theory of Justice* (Cambridge MA: Harvard University Press, 1999 [1971]), pp. 319ff.

[38] Varoufakis ('Europe's Left after Brexit').

ways to distinguish principled disobedience from outwardly similar forms of exceptionalism.[39]

Agents intending to work outside the existing framework of rules need to invoke a non-procedural authority in support of their actions. In the case of established political elites, this often takes the form of a claim to technical expertise and effectiveness. As Chapter 5 showed, executives' powers may present themselves as having special insight into the situation at hand and the capacity to respond swiftly and effectively. They may seek thereby to authorize themselves to go beyond the rules to preserve the *spirit* of the rules. But for those seeking to contest such moves 'from below', authority claims of this kind will typically be unavailable. The kind of knowledge certified as expertise may be exactly what they wish to challenge, and speedy fire-fighting actions the kind likely to reproduce the defects of the status quo.

An alternative source of authority—one that established powers are often badly able to claim—is the readiness to submit actions to public approval. Agents of disobedience who demonstrate this willingness have one basis on which to authorize their departures from the constraints of existing politico-legal frameworks. By inviting the involvement and evaluation of a wider public, they acknowledge their actions must not be unchecked. In effect, they substitute the constraint of public approval for the constraint offered by rules. They take an important step thereby towards distinguishing their actions from arbitrary or self-interested discretion.

There are various ways such authority can be claimed. The willingness to contest an election or referendum with reference to the proposed disobedience is one. The 'No' of the Greek referendum of 5 July 2015 offered the Syriza-led government further backing for a strategy of non-compliance with the demands of the Eurozone's crisis management. Although Syriza ultimately did little to build on the vote, it was a form of authorization nonetheless. Executive discretion in the Euro crisis, by contrast, was generally the action of elites acting independently of public control mechanisms, and to the extent that they had an electoral mandate at all it was one weakly linked to the measures advanced. One may also highlight forms of recourse to public approval less open to the charge of plebiscitarianism, based on seeking popular involvement in the formulation of policy, e.g. through participatory forms of party membership. In addition to ex ante forms of authorization for disobedience, there are also ex post measures—e.g. the calling of elections—by which a readiness to submit actions to public approval can be shown.[40]

[39] For recent interventions: Scheuerman (*Key Concepts*). Celikates ('Rethinking Civil Disobedience'); Robin Celikates, 'Democratizing Civil Disobedience', *Philosophy and Social Criticism* 42 (10) (2016), pp. 982–94.

[40] In discussions of civil disobedience, the principled nature of the act is often said to be signalled by willingness to accept sanction. Equivalent ex post actions may be available to disobedient actors working through institutions, including resignations of office.

On its own, this seems a promising but insufficient basis for principled disobedience. In the European context, the approval that can be sought this way is likely to correspond to national or sub-national institutions. Though such solicitations of approval mark a step away from an arbitrary or self-interested exercise of discretion, they remain compatible with what from a broader perspective remain parochial ends.[41] When Hungary's ruling Fidesz party took aim at EU rules and proposals on immigration and asylum in 2016, it styled this as an elected national government speaking up for the views of its people, with a national referendum intended to underline this.[42] To be sure, some rather dubious means may be used to manufacture such opinion, and the right-wing populist outlook is generally wary of genuine participation and the party structures that enable it.[43] But still, some kind of recourse to public evaluation may characterize a wide variety of forms of disobedience.

A second criterion of legitimate transgression that builds on this can be expressed as a commitment to generalizable justification. Of interest here is when acts of disobedience against the existing regime are coupled with an account of the normative goals they are intended to serve.

We have seen the prominence of ideas of necessity in rationalizing the manoeuvres of emergency rule. By the standards of political justification, this type of framing is weak: it obscures the value choices such actions entail, and denies the possibility of viable alternatives. To be sure, ideas of last resort are a staple of thinking about civil disobedience. But what gives such transgressions a principled character is when they are paired with an account of what is valuable and ought to be preserved—of how a given course of action not only allows the association's survival but its survival in a good form.

Notable in the claims of Syriza in spring 2015,[44] and replicated in the programmatic statements of successor movements, was the emphasis not only on highlighting these normative stakes but placing them in a pan-European perspective.[45] Actions were traced out to the normative priorities of Europe as a whole: the focus was on *generalizable* ends. Consistent with this, disobedience can be deemed principled to the extent it is pursued on behalf of a wider whole, not just an effort

[41] This is why seeking public approval does not entail leaving the disobedience paradigm for that of constituent power—contra Markus Patberg, 'Destituent Power in the European Union: On the Limits of a Negativistic Logic of Constitutional Politics', *Journal of International Political Theory* 15 (1) (2019).

[42] See the referendum of 2 October 2016.

[43] Müller (*What is Populism?*), pp. 29, 35.

[44] See e.g. an early effort to fix the meaning of the result of the 5 July referendum: https://yanisvaroufakis.eu/2015/07/06/our-no-is-a-majestic-big-yes-to-a-democratic-rational-europe/

[45] 'To be progressive and constructive, we must accompany disobedience by counter-proposals fully outlining alternative policies or directives to those we disobey. These alternative policies or directives must, in addition, be *universalisable* (in the Kantian sense of being the policies one should want to see adopted, at once, throughout Europe). In other words, our disobedience does not qualify as constructive if the alternatives we propose are based on the NIMBY (Not In My Back Yard) syndrome, or are of the type that, if adopted in every country, some parts of Europe will suffer.' (https://diem25.org/constructive-disobedience/)

to extricate particular agents from particular constraints.[46] In the EU context, though appeals to the welfare of particularist groupings such as country and region will typically feature, especially when approval through local institutions is sought, the challenge is to locate these within an argument about the larger interest.

There is a contrast to be made here with populist Far Right forms of disobedience, insofar as the latter treat as self-evident and incontestable the truth of the claims advanced.[47] They typically seek not to ignite debate and persuade but to articulate what 'everyone' is said already to know. The appeal in the first instance is to authenticity rather than justification. Moreover, the goals pursued tend to be explicitly *non*-generalizable, as when Fidesz's challenge to migration rules was paired with the expectation that other countries would adhere to them. The disobedience agenda was decidedly national in focus.

A third criterion of principled disobedience deserving emphasis is a commitment to achieving a genuine constitutional order. Considerations of this sort are often invoked in discussions of civil disobedience by reference to the singular importance of 'very highest respect for the law'.[48] Disobedience is presented as legitimate to the extent it is wrapped up in larger statements and actions of fidelity. Often the understanding is narrow, as when the disobedient agent is expected to endorse the existing legal order largely as it stands, rejecting only those specific rules it chooses to transgress. In the case of the EU, where constitutional foundations are imperfect at the best of times, respect for the law may be better understood as a broader commitment to a constitutional order.

A commitment to the rule of law is typically voiced 'from above' by those seeking to normalize the emergency regime post hoc. Yet it clearly matters how far such stated commitments are backed by actions. However energetic has been the production of new rules in the EU in recent years, it scarcely demonstrates commitment to a constitutional framework. Not only do these rules leave significant scope for discretion in the hands of powerful actors, but their detailed stipulations are likely to fare badly as conditions evolve, leaving them vulnerable to further acts of far-reaching revision. To quote Joerges again: 'de-legalization' is the tendency expressed in this recourse to binding controls.[49]

Commitment to achieving a genuine constitutional order is also where many Far Right forms of disobedience do badly. The non-pluralist tendency typically finds expression in an impatience with procedural norms, with these regarded less as something to institutionalize than as a threat to identity, agency, and commonsense. While pragmatism may encourage Far Right politicians to avoid illegality

[46] Cf. Scheuerman ('Recent Theories of Civil Disobedience'), p. 7.
[47] Müller (*What is Populism?*), pp. 20, 31.
[48] For critical discussion: Scheuerman (*Key Concepts*), pp. 147ff.
[49] Joerges in Chalmers et al. (*End of the Eurocrats' Dream*).

where possible, and even to support constitutional measures where these can serve their agenda, the tenor of their practice and discourse is generally one of scepticism towards political and legal constraints.[50]

What does it mean to show commitment to a constitutional order in the EU context? Aside from stating this goal and advocating measures that serve it, there is a distinctive opportunity that transnational politics affords to groups pursuing institutional disobedience. Their respect for law and process at the transnational level can be gauged in some measure from their conduct at the national level.[51] A party is typically a disciplined force compared to the civil disobedience of crowds, and one that has entered government or parliament at the national level will have had opportunity to show its readiness to adhere to constitutional requirements. In doing so it may assuage concerns that its rule-breaking is arbitrary and liable to be turned back on its own citizens. A commitment to the *principle* of a constitutional order—though not, of course, to all the particulars of the EU framework—is something agents of this kind can credibly indicate, however disobedient their actions transnationally may be.

Exceptional Exceptionalism

These considerations on the legitimacy of disobedience give a vantage-point for reflecting more generally on the legitimacy of transnational emergency rule. The argument has been that certain forms of disobedience can claim a normative foundation unavailable to the emergency actions of established executive powers. The latter generally involve transformative steps taken with little effort to seek public approval, with values obscured by discourses of necessity, and with indifference to a genuine constitutional order. What then can one say of the kind of world that disobedience aims to create? Should the goal be to exclude the possibility of emergency rule altogether, or somehow to circumscribe it such that it takes an acceptable form? Is the exceptionalism of the disobedient for the sake of transition to an order beyond exceptionalism?

Whether there can ever be legitimate recourse to emergency rule is a long-standing question. We have noted the difficulties of settling it by appeal to the nature of the external conditions. The intuition that transgressing basic rules and norms is reasonable to the extent that necessity demands it, though it may alert one to flagrant abuses, will be little guide in more complex cases. The

[50] Kim Lane Scheppele, 'Autocratic Legalism', *University of Chicago Law Review* 85 (2) (2018), pp. 545–83; Müller (*What is Populism?*), pp. 50, 63.

[51] Only in part: as noted earlier, the insertion of economic rules into the national constitutions of Eurozone states means national and transnational rule-breaking will be increasingly difficult to separate.

circumstances at hand are rarely so clear. To pretend they are is to accentuate the depoliticizing tendencies of emergency rule. Some of the tensions here are already visible in Schmitt's work: though often he wrote as though concerned with challenges to the established order that are unmistakeably existential, the perspectival aspect is implied with the notion that sovereignty is expressed in *deciding* when a situation is exceptional.

The criterion of necessity leaves too much to hang on supposedly objective features of the conditions at hand, or—as is likely to be the practice—on the judgement of those claiming the authority to interpret them. Other criteria, potentially less authoritarian, are also commonly invoked in discussions of exceptionalism. One concerns the extent to which extreme situations are *foreseeable*. The suggestion is that unconventional responses are less legitimate to the extent that the challenges at hand could be anticipated.[52] Again, such a criterion may have some use when applied negatively to criticize over-willingness to resort to exceptionalism. What can be foreseen—'emergencies waiting to happen'—should be prepared for, such that when they occur they can be handled in a regular fashion. Here is the basis of an argument for the simplification and constitutionalization of the EU, to which the next chapter returns. If one driver of emergency rule is the diffusion of power across multiple institutions and their uncodified relations, these should be addressed in anticipation of future challenges.

Moreover, what is not only foreseeable but *deliberately sought* would seem to provide no warrant for exceptionalism—a pertinent point in the EU. It is sometimes said, including by its officials, that EU policy-making proceeds in the full knowledge that its deficiencies will prompt recourse to emergency measures in future. If the Euro was created in awareness that it could bring about the exceptional circumstances required to advance economic union, the warrant for emergency rule would seem especially thin. Here is another respect in which the exceptionalism of the disobedient is different: it generally deals with problems not of its protagonists' own making.

But if emergency methods applied to foreseeable or contrived challenges are particularly dubious, clearly it does not follow that there are situations where such methods are legitimate. As with necessity, the application of such criteria is never mechanical: it depends on a judgement call, one likely to be in the hands of already-dominant executives easily swayed by competing motives. This is the danger of endorsing criteria that can act positively as markers of legitimate emergency rule: they may be invoked too readily by those in a position of relative strength. All such criteria invite their abuse by those with ideological or power-building agendas. One should be sceptical of seeking criteria that can legitimize large-scale discretion from above.

[52] Cf. Sorrell (*Emergencies and Politics*), p. 5.

In a highly imperfect order such as the EU, the capacity for creative actions that move beyond existing structures is undoubtedly important. No amount of reform achieved through practices of disobedience is likely to do away with it. But the tendency to pursue change by invoking the licence of emergency conditions is one to be resisted. Executive discretion in the transnational context tends to yield lasting transformations, deserving the considered and inclusive deliberation that the politics of emergency precludes. Even principled disobedience is quite imperfect in this sense. Though it may help challenge past abuses and initiate a deliberative process, more inclusive and procedural forms of participation seem indispensable.

If emergency rule is a problematic mode of polity-building, is there not a place for it in extreme conditions as a way to maintain order? What of the danger of leaving the EU at risk of paralysis in times of genuine crisis: are there not circumstances when executives *should* be able to act with heightened discretion, unencumbered by the usual checks? One might imagine for instance a process whereby an EU state of emergency could be formally triggered by European and national parliaments, vesting exceptional powers in an agent of their choice—the Commission, ECB, Council, or some combination. If reassessed at regular intervals, such a delegation of power might seem to augment the capacity to act while ensuring claims to necessity were carefully weighed.[53]

Again, the concern must be that such mechanisms would be abused to institute measures that, even if formally temporary, are difficult to reverse, likely to set precedents, and laden with unstated value implications. Given the number of veto players in the EU, decisions once taken are hard to undo. Giving executives a strictly limited period in which to act would arguably merely encourage hasty and assertive interventions. In an age of *governance* especially, when executives are weakly constrained from within and closely tied to socio-economic power, the costs of emergency rule seem too high.

Moreover, the structure of the transnational realm should make us more reluctant to conceive the kinds of threat that might warrant such actions. The break-up of transnational policy regimes, even the EU itself, need not herald a return to the state of nature, as there are the institutions of the state to fall back on. An existential threat to the one need not be such to the other. While the demise of a transnational order may well have negative implications for cross-border peace and security, it certainly need not lead directly to a condition of anarchy.

[53] Such a model might draw on Bruce Ackerman's notion of a 'super-majoritarian escalator', whereby the grant of emergency powers requires ever-higher thresholds of parliamentary support as time passes (Bruce Ackerman, *Before the Next Attack: Preserving Civil Liberties in an Age of Terrorism* (New Haven: Yale University Press, 2006). For a proposal to restrain ECB-led emergency rule in similar fashion: Jens Van 't Klooster, 'Democracy and the European Central Bank's Emergency Powers' *Midwest Studies in Philosophy* 42 (1) (2018); cf. Kenneth Dyson, 'Sworn to Grim Necessity? Imperfections of European Economic Governance, Normative Political Theory, and Supreme Emergency', *Journal of European Integration* 35 (3) (2013), pp. 207–22.

Emergency rule, one may conclude, is something to oppose rather than seek criteria of legitimacy and provision for. The constitutional order for which the principled-disobedient strive is one that should attempt to do away with it. Disobedience should be understood as aiming to correct the failings of the existing order rather than to consolidate a predilection for executive discretion.

The next chapter examines how emergency rule might be constrained in the longer term. The focus here has been on challenging its legacy rather than preventing it, but the same party-based forms may be relevant to both. A collective that ties its leaders into a larger structure and demands that actions be reconciled with programmatic commitments may go some way to compensating for the limitations of other restraints. It may also be a way to cultivate public vigilance. Where institutions are instead populated by relatively detached officials, exceptionalism is that much easier to carry through.

Conclusion

The contestation of emergency measures faces numerous challenges, magnified in the transnational context. Efforts to unpick such measures by ordinary legislative means are likely to founder on the reluctance of key agents and their vetoing capacity, while judicial responses face distinct problems of legal and epistemic authority. A strategy of malicious compliance intended to sharpen the regime's contradictions is meanwhile quite uncertain in the outcomes it leads to. This leaves a politics of principled disobedience as one of the few ways to resist the normalization of measures introduced under the sign of emergency.

Disobedience is primarily about influencing opinion, putting the legitimacy of current arrangements in question. Simply obstructing the consolidation of an unjust order is a sizeable objective. More challenging is for disobedience to exert a deeper influence on opinion by finding recognition as a principled stance. The transnational setting raises some distinctive challenges, given the uncertain constituency for disobedient actions and the insulation of decision-making from popular interference. As the Greek episode in 2015 suggests, such forms of defiance must contend not only with the possibility that they meet with indifference, but with efforts to portray them as unprincipled recalcitrance, as one more reason to insist on the rules.

The opportunity to wage disobedience through institutions is one of the intriguing features of the transnational realm. Rethinking the agents as intermediary associations like parties is one way to acknowledge this potential. Though its disobedience in 2015 was ultimately more threatened than performed, Syriza could credibly claim to have sought popular backing for a policy of non-compliance, to have offered generalizable justification for such actions, and to have expressed its underlying commitment to a constitutional order. The normative basis for its challenge deserves recognition, particularly given the potential for ostensibly similar forms of disobedience to emerge from the Far Right. Its slim practical effects suggest either that the EU is beyond reform—a possibility that disobedience must always contend with—or that efforts to reform it depend heavily on a favourable arrangement of political forces cross-nationally.

8
Structuring Partisan Power

In the crisis politics of the 2010s, Europe's executive branches used extreme conditions as a licence to reshape the EU, consolidating certain key policies and the wider socio-economic order they serve. Disobedience towards the structures arising is one plausible rejoinder, but it marks an incomplete break with the logic of exception, as a form of exceptionalism 'from below'. In this penultimate chapter, I want to examine the longer-term prospects for containing emergency politics in this setting. On the assumption that it remains a persistent possibility, I explore how an effective response to emergency rule might look, one that both wards off its exercise and counters some of the harms it does when it occurs.

My focus remains on the politics of emergency as it finds expression in the EU. I am interested in how constraints might take shape in this setting rather than how they might emerge through exiting or dismantling it. Some may be tempted to find in the patterns of emergency rule examined an argument for relocating authority back to a world of sovereign states, but this is not the possibility pursued here. As the large contemporary literature on security measures post-9/11 makes clear, the contemporary state is hardly a refuge. Though it may be less vulnerable to emergency politics in some regards, indirectly it is prone to many of the same dynamics. More importantly, as the Brexit scenario may suggest, efforts to extricate from the transnational sphere can themselves be the motor of executive empowerment, and may eat into democracy and solidarity in numerous ways. Even if the national alternative were in principle superior, reaching it can undermine the very qualities that were sought in it.

In the large body of scholarship on exceptionalism, discussions of how emergency rule may be constrained tend to divide between legal and political approaches.[1] The following is oriented to the latter. Some of the reasons for scepticism towards legal responses in the transnational context have already been examined, and pick up on wider concerns about judicial deference in high-stakes moments. Without further questioning the role of courts, here the focus is elsewhere.

Within the tradition that seeks *political* constraints on emergency rule, we may distinguish those offering primarily institutional solutions from those prioritizing

[1] See Victor V. Ramraj, 'Introduction', in Ramraj (ed.), *Emergencies and the Limits of Legality* (Cambridge: Cambridge University Press, 2008), esp. pp. 15–16. For an authoritative court-centred approach: Dyzenhaus (*Constitution of Law*).

Politics of Last Resort: Governing by Emergency in the European Union. Jonathan White, Oxford University Press (2020).
DOI: 10.1093/oso/9780198791720.001.0001

public mobilization. A classic example of the first would be Alexander Hamilton's arguments for individualizing executive power so as to make it more easily held to account. For Hamilton, the diffusion of power across multiple executive agents was always liable to loosen the constraints:

> [T]he plurality of the executive tends to deprive the people of the two greatest securities they can have for the faithful exercise of any delegated power, *first*, the restraints of public opinion, which lose their efficacy, as well on account of the division of the censure attendant on bad measures among a number as on account of the uncertainty on whom it ought to fall; and, *second*, the opportunity of discovering with facility and clearness the misconduct of the persons they trust, in order either to their removal from office or to the actual punishment in cases which admit of it.[2]

Authority was to be vested in the single person rather than the group. As the example of the Roman decemvirs illustrated, 'a council to a magistrate, who is himself responsible for what he does, are generally nothing better than a clog upon his good intentions, are often the instruments of accomplices of his bad, and are almost always a cloak to his faults'.[3] Though Hamilton was more than willing to grant that executive power might need great discretion in extreme moments, he insisted that its institutional structure should be uncomplicated.

Such arguments seem all the more pertinent in the transnational context, where the structure of executive power is far more complex and diffuse than anything Hamilton foresaw. We are not just talking about a council of persons rather than an individual, but multiple 'councils' spanning the national and supranational spheres, engaged in a precarious process of coordination. Even if, as modern democrats, we may be more sceptical than Hamilton towards placing executive power in the hands of an individual, there would seem to be great merit in institutional simplification and codification. If weakly defined constitutional relations are one of the structural vulnerabilities of the transnational sphere to emergency discretion, and tendencies towards shape-shifting and de-institutionalization one of its more pernicious outcomes, then efforts to ensure institutions have clearly defined roles they adhere to, and defined channels of communication and coordination, are surely desirable.[4]

A more integrated transnational executive would arguably be less prone to the politics of emergency, as it would minimize the persistent problem of uniting disparate agents behind a common decision. In the terms of Chapter 6, it could

[2] Hamilton, *Federalist Papers* (no. 70), p. 348.

[3] Hamilton, *Federalist Papers* (no. 70), pp. 349/50.

[4] Cf. Chapter 4; for detailed proposals on central banking: Paul Tucker, *Unelected Power: The Quest for Legitimacy in Central Banking and the Regulatory State* (Princeton: Princeton University Press, 2018), ch. 23, esp. p. 523.

swap *agency* for *doing*, building a deeper basis of public support. Such a reconfigured executive would be better equipped to pursue wider programmes of socio-economic change, addressing thereby some of the exogenous drivers of exceptionalism. Even if it did none of these things, instead resorting persistently to extraordinary methods, such a reconfigured executive would be a more visible target for critical opinion. As Hamilton foresaw, the attribution of responsibility would be improved. Extending his reasoning beyond his own arguments, one might conclude that the best response to the transnational politics of emergency is the construction of a powerful and unitary transnational executive, embedded in institutions of transnational democracy.

Hamilton though was writing in a constitutional moment, when basic questions of design were up for negotiation. There was the prospect of conceiving institutional arrangements that served not just to sediment the preferences of status quo powers but that could plausibly claim to express constituent power. Developing blueprints to constrain executive discretion makes sense under such conditions of constitutional foundation, or when these can be expected.

Is the EU in a constitutional moment? This is a matter for tomorrow's historians. At the time of writing one is inclined to think not, though certainly there is an expanding array of movements calling for conventions and constituent assemblies. As the last chapter indicated, principled challenges to the status quo derive part of their motivation from the hope of bringing these closer. Possibly the conditions will emerge in which there is receptiveness to blueprints amongst those with constitutional authority. Possibly one should exercise the political imagination anyway, constitutional moment or not. Some of the more interesting contemporary scholarship on the political theory of the EU analyses how an exercise of constituent power might look, as well as the good reasons to seek it.[5]

But in developing an account of the political restraint of emergency rule I prefer the second route, focusing less on matters of constitutional redesign and more on political mobilization within broadly existing structures.[6] This is partly to avoid relying on opportunities for overhaul that are rare. Most of the time, one can expect established powers to resist such changes, even if, as legitimacy-building moves, they are in their enlightened interest. It is partly also because constitutional designs should be the outcome of a participative process. The deliberation of a constituent assembly has its own dynamic for which individual reflections are no substitute.

[5] Patberg ('Destituent Power'); Fabio Wolkenstein, 'Transnational Partisanship and Networked Constituent Power in the EU', *LEQS Paper* 138 (2018); Peter Niesen, 'Reframing Civil Disobedience: Constituent Power as a Language of Transnational Protest', *Journal of International Political Theory* 15 (1) (2019), pp. 31–48.

[6] In the institutional tradition, see Ackerman (*Before the Next Attack*). Posner and Vermeule (*Executive Unbound*) also emphasize political rather than legal constraints on executive power: their suggestion that plebiscitary mechanisms backed by media commentary are adequate to this task seems optimistic though.

But most of all it is because constitutional reforms need those who can press for them and later defend them. Even where changes are successfully introduced, the resulting order needs to be maintained, and some degree of mobilization is likely to be required. As the experience of constitutional settings such as the US suggests, institutional mechanisms alone do not ward off exceptionalism in the executive. What is true in the state setting is likely to be all the more true in the transnational sphere, where leaping the bounds of one institutional regime to establish another is always on the cards. More agential constraints on power seem indispensable.

In the sections that follow I map out an account of the constraint of emergency rule centred on practices of partisanship. Picking up one of the book's earlier arguments, that tendencies towards exceptionalism are a symptom of the hollowing of parties and the rise of an ethos of governance, I examine how restraining these tendencies may depend on the renewal and extension of partisanship. By controlling executive institutions from within, and especially by challenging their actions from without, parties may be in a position to contain emergency politics both ex ante and ex post, and also to counter some of its negative outcomes when it occurs. They can make it harder for established powers to justify their exceptionalism and to assume resistance can be discounted. Unfolding this argument is the task of the first section.

Later I explore how binding executive power in this way is ultimately about cultivating a certain temporal structure of politics. It is about facilitating the coordination of critical opinion and the legitimacy of dissent in high-stakes moments, for which periodic modes of contestation have a particular significance. Whereas key EU decisions in recent years have often been challenged on an ad hoc basis, in the form of street protests, strikes, and referenda, more lasting forms of opposition are likely to depend on the consolidation of cycles of contestation. As the final section suggests, although these remain weak on the European scale, in principle they can be developed further by the creative use of electoral and non-electoral channels. The contest of national elections by transnational parties looks particularly interesting in this regard. Though such practices would not signal an end to the transnational politics of emergency, they would strengthen the opposition that can be marshalled against it, as well as reinforce some of the things it weakens.

Partisan Constraints on Emergency Rule

Emergency politics in the EU has much to do with the rise of a managerial ethos of governance—this was Chapter 3's argument. The exercise of far-reaching discretion by largely autonomous executive agents expresses the rise of an outlook prizing efficiency in problem-solving with limited regard for legal and political norms. Governance sees executive agents define themselves by pragmatism and flexibility. Uncertain of their ability to connect with mass publics by reference to shared ideologies and normative commitments, yet inclined to press ahead with policy reforms nonetheless, they rely on discourses and actions of last resort to ward off potential resistance.

If this is broadly correct, one way emergency politics may be restrained involves displacing this managerial ethos from the corridors of power. Challenging tendencies to exceptionalism at the root entails changing the governing outlook from within, repopulating executive institutions with those that put political commitments to the fore rather than the kind of instrumentalism that spills into exceptionalism. In the EU context, the institution most susceptible to such reorientation looks to be the European Council, given its connection to electoral politics. Political parties that define themselves according to a clear set of principles and organize on a participative basis are one candidate agent for leading such change.

Consider some general features of the party as a political form.[7] As an agent expected to define itself by principled ends, a party has no necessary dependence on discourses of emergency to shore up its authority with mass publics. Though it may sometimes have reason to invoke these, it is not structurally drawn to them in the way the governance model implies. A party can eschew purely technical rationales for action and discourses of necessity, grounding itself instead in normative commitments intended to be generalizable and widely intelligible. By developing a comprehensive ideology, it can put political challenges in context, relating them to a broader narrative rather than treating them as a succession of discrete problems. It can pursue an interconnected and historical perspective, quite different from the emergency outlook in which socio-economic problems are treated as suddenly-appearing and exceptional.

Organizationally, parties offer resources for binding elites to a larger structure. When configured on a participative model, they allow the collectivization of decision-making. Mechanisms of intra-party democracy are one way the discretion of the few may be constrained.[8] More politically engaged than citizens at large, partisans are well placed to press leaders to act in line with shared principles. One of the shortcomings of many political approaches to the constraint of emergency rule is that they place the burden of scrutinizing executives on a broad public. The mainstream media or 'the people' in general may be too quick to accept political exceptionalism. In Carl Friedrich's words, 'behind all these procedural devices there must stand an alert people, a real constituent power, determined to see to it that these limitations [on dictatorship] are effectively utilised'.[9] Plausible as this may be, peoples need those who can mobilize

[7] For further discussion: Jonathan White and Lea Ypi, *The Meaning of Partisanship* (Oxford: Oxford University Press, 2016).

[8] Wolkenstein ('Deliberative Model').

[9] Friedrich (*Constitutional Government*), p. 581; cf. Gross in Ramraj (*Emergencies and the Limits*), p. 64; Tushnet in Ramraj (*Emergencies and the Limits*), p. 151. For apt remarks on the place of ideology

them. Engaged partisans are potentially an important intermediary—in the national context too, but especially in the weak public spheres of the transnational realm.

Mechanisms of intra-party democracy are also a way to insulate leaders from external pressures, including from non-state securitizing agents who encourage resort to emergency measures. They are a means to uphold normative projects in the face of adversity—to undergird political *commitment*.[10] Certainly, determined elites can slip the bounds of party organization just like they can slip the wider norms and procedures of the polity, but in the structure of the partisan association there are important resources for reinforcing constraints on their discretion.

In much of the critical scholarship on emergency rule, there is a tendency to treat executive authority as intrinsically susceptible to authoritarian impulses. This may be because those writing on the topic are often liberal scholars concerned primarily with the preservation of individual rights rather than norms of collective self-determination. One of the advantages of reflecting on partisanship is that it cautions against undifferentiated scepticism towards executive power. The latter becomes, I suggest, especially prone to exceptionalism under conditions where parties are weak. Strengthening them is one route towards the restraint of emergency rule.

Yet we saw in the previous chapter the difficulties that face committed partisans in imprinting themselves on executive power. Principled disobedience is something to which agents in a minority position can plausibly aim, but even this may depend on a favourable cross-national alignment. A deeper effort to reorient the exercise of executive power across the EU would seem to require the emergence of strong parties across a much wider space. The proliferation of institutions in the transnational sphere, though it creates opportunities for the occasional radical party to gain a foothold, makes it that much harder for a larger overhaul. And though the Council has been one of the main sites of emergency discretion in recent years, it is flanked by many others that are deliberately insulated from partisan influence, notably the Commission, ECB, and relevant externals such as the IMF.

If one is willing to envisage major institutional innovations, certainly there are some that might strengthen the capacity of parties to impose themselves on executive power. Further linking the composition of the Commission to the outcome of European-Parliamentary elections, e.g. by requiring that Commissioners be MEPs, would be one. Rather than individuals selected on largely technocratic criteria, incumbents would then be the representatives of Europe-wide

and political action in constraining emergency rule: Campbell in Ramraj (*Emergencies and the Limits*), p. 220; Scheuerman in Ramraj (*Emergencies and the Limits*), pp. 271–2.

[10] White and Ypi (*Meaning of Partisanship*), ch. 4.

parties.[11] Coupled with greater Commission powers vis-à-vis the Council and ECB, and the 'de-constitutionalization' of the policies of the *acquis communautaire,* one would see something closer to a unitary elected government of the EU and a target for party control. Again, constraining emergency rule in the EU seems to go hand-in-hand with simplifying the structure of executive power. Combined with a sharper codification of the relations between national and supranational institutions, one would have a clearer delineation of the locus of power and thus a clearer basis for public identification. The more the transnational sphere comes to resemble the state model, in other words, the more one could expect executive power to be susceptible to constraint.

Should a favourable alignment of parties across national governments arise, it would create a window of opportunity to push for changes of this kind. The history of transnational partisanship is not without such transformative episodes.[12] But if emergency rule is to be constrained by parties in something like the existing structure of the EU-sphere, in the first instance this is likely to be from a position outside the main executive institutions. We need to explore the difficult case, in other words, of a transnational partisanship of opposition—one organizing itself not against a well-defined government but a diffuse network of executive power.[13]

On a transnational scale, partisanship involves the ongoing collaboration of cross-border communities of the like-minded, organized as supranational parties or alliances of national parties.[14] To be an effective restraint on EU emergency rule, such activities would need to address decision-making in multiple locations, building the coordination of critical opinion across them. In some cases, contesting the executive usurpation of power would be the very rationale for cooperation. In others, cooperation would be organized around different themes—e.g. structural injustice, or models of economy—and would have emergency rule in its sights only to the extent it interferes with these. In all cases, the task would be to articulate transnational positions of principle, ensuring executive power is the subject of critical scrutiny and knows itself to be such.

The point of opposition is not just to obstruct. It is not just a way of saying 'no', but of constructing viable alternatives.[15] This is why the transnational dimension matters. To the extent that opposition means blocking, some degree of it can be

[11] For an argument for extending the principle of parliamentary nomination well beyond the *Spitzenkandidaten* model: Joseph Lacey, *Centripetal Democracy: Democratic Legitimacy and Political Identity in Belgium, Switzerland and the European Union* (Oxford: Oxford University Press, 2017), pp. 221–3.

[12] White and Ypi (*Meaning of Partisanship*), ch. 9; Karl Magnus Johansson, 'The Role of Europarties in EU Treaty Reform: Theory and Practice', *Acta Politica* 52 (3) (2017), pp. 286–305.

[13] This possibility is briefly outlined in White and Ypi (*Meaning of Partisanship*, ch. 9), though with an emphasis on the obstacles.

[14] White and Ypi (*Meaning of Partisanship*), ch. 9.

[15] On this theme: Klein (*No is Not Enough*).

marshalled at a purely national level, and Far Right parties have arguably already had success in challenging EU emergency politics in this way. But to the extent that it is about cultivating alternative visions, able to prepare a deeper reorientation of political and economic structures, it needs coordination on a transnational scale.

If the EU's origins as a purposive association are one of its deep vulnerabilities to emergency rule, inviting far-reaching discretion in the name of its policies, then challenging executive power involves challenging what those purposes entail, what can be justified in their pursuit, and offering competing views. It requires contesting these purposes so they can less easily be invoked to justify whatever established agents have a will to do, and so it becomes harder to constitutionalize a certain conception.[16] Opposition involves demanding from those in authority principled justifications for their actions, rebuilding the sense of agency that is disavowed through discourses of necessity. It entails arming people with historical perspective, so that ideas of exceptional circumstance are less easily deployed to waive democratic norms. Most basically, it entails identifying and publicizing *who* the agents of executive power actually are—identifying, that is, who rules.

An obvious problem is that emergency discretion tends to be motivated exactly by the aim to neutralize opposition. The stronger the constraints, arguably the stronger the temptation to resort to the practice. But as earlier chapters have indicated, such dynamics are well advanced in the EU. Stronger dissent is not going to tip the balance decisively in favour of emergency politics, as in the current configuration that balance is already lost. The question is how best to scrutinize and challenge the practices of emergency rule, on the reasonable assumption they are unlikely to recede spontaneously.

Are there parties capable of playing such a role in today's Europe? Given the long-standing complicity of many in the EU's turn towards governance, one may be sceptical more generally of their capacity to contest power. Political scientists diagnose this as the age of 'cartel' parties,[17] and the politics of emergency is one consequence of their unwillingness to compete around contending ideologies. If large numbers of such organizations have become devoid of *partisanship*, at the leadership level especially, how plausible is it to suppose that others can act as meaningful constraints on the executive?

Reasons to suppose they might do so lie in some of the tendencies described in this book. A central element of the cartelization thesis is that mass publics are increasingly indifferent to political affairs and have withdrawn into private concerns.[18] The mobilizations of recent years against established political authorities

[16] Cf. Grimm (*Constitution of European Democracy*) on over-constitutionalization.

[17] Richard Katz and Peter Mair (2009), 'The Cartel Party Thesis', *Perspectives on Politics* 7 (4) (2009).

[18] Mair (*Ruling the Void*), pp. 16–17.

suggest this depoliticization thesis is overstated. Those contesting practices of governance in the name of restoring political agency have tended to do well, and not always on the basis of an exclusionary nationalism.[19] The appeal of 'populism' should not be equated with its least savoury aspects: it is testament also to an enduring political receptiveness, at least when leaders ditch the language of necessity. Demand for a more forthright politics is not lacking, but the challenge is how to channel it constructively.[20] A politics based on clear statements of principle and enduring commitment may still be one route to electoral success: 'cartel parties' need not be the only ones to prosper.

Coordinating partisanship on a transnational scale is clearly the deeper challenge. Certainly, some of the arguments traditionally raised against this in the EU—for instance, that no salient and contentious matters are decided at a European level—look decidedly obsolete in the context of the emergency politics of the 2010s. The notion that political ideologies inevitably fail to intersect across borders—that there is little continuity in the structure of political division[21]—looks dubious too. The shared experience of EU crisis politics is itself one potential source of cross-border unity of purpose. But beyond such affinities, we need to look more closely at how transnational practices of contestation can emerge, and emerge in a form that speaks to the concerns in question. What resources can be drawn on, what platforms are available?

The Temporal Construction of Political Opposition

One of the familiar but under-analysed features of modern democratic institutions is the *periodicity* by which they operate. They follow a temporal structure based on alternation and repetition. The contestation of decision-making is organized such that opportunities repeat, and repeat with some measure of regularity. In the electoral sphere, time rules set an outer range to the intervals between elections to office, establishing a pattern of voting that—if not always evenly spaced—has a recognizably cyclical structure. In the legislative sphere, timetables of assembly and debate guide the actions of representatives, producing a pattern of activity that is cyclical. Electoral and parliamentary cycles are just some of the institutional structures in which modern democracy gives expression to the rhythmic form.

[19] See e.g. the post-2015 experience of the British Labour Party under Corbyn, and Syriza in spring 2015.

[20] Carlo Invernizzi Accetti and Fabio Wolkenstein, 'The Crisis of Party Democracy, Cognitive Mobilization, and the Case for Making Parties More Deliberative', *American Political Science Review* 111 (1) (2017), pp. 97–109; Lise Herman, 'Democratic Partisanship: From Theoretical Ideal to Empirical Standard', *American Political Science Review* 111 (4) (2017), pp. 738–54.

[21] White and Ypi (*Meaning of Partisanship*), ch. 9.

In the small body of research on the topic, the significance of periodicity tends to be located in the check it presents on authoritarian rule. Institutions on the periodic model express the idea that government is *pro tempore*, that power is to be held for a limited period.[22] Claus Offe speaks of the 'the constitutive temporal limitation of power through the periodicity of elections'.[23] By intermittently requiring executive institutions to re-staff themselves with different officials, the periodic structures of electoral politics help separate the offices of rule from the persons who occupy them. In this sense, such rhythms become a symbol of and a contribution to accountable rule. Hereditary monarchy, as an essentially continuous rather than periodic principle of authority, would be one familiar point of contrast, along with technocratic institutions and judicial authorities.

The argument I propose is that the periodicity of institutions is conducive to the emergence of political opposition. For its consolidation in the national context, such a temporal frame was important. Building such agency at a transnational level will depend likewise on the creative appropriation of such rhythms. Partisan associations, just like a whole range of collaborative endeavours, are bound up in the periodic form.[24] We can see this better if we reflect theoretically on some of the features critical to its effectiveness.

The contestation of executive decision-making depends on the capacity of critical opinion to organize itself. It depends on dissenters being able to concentrate their resources and constitute themselves as some type of collective agent, achieving a power greater than that of individuals. This basic challenge of coordination is one to which institutions on a periodic model respond. Periodicity first and foremost allows expectations about the future to form, on the basis of which joint action between agents becomes possible.

[22] See Juan J. Linz, 'Democracy's Time Constraints', *International Political Science Review* 19 (1) (1998), pp. 19–37; Andreas Schedler and Javier Santiso, 'Democracy and Time: An Invitation', *International Political Science Review* 19 (1) (1998), pp. 5–18; Henning Laux and Hartmut Rosa, 'Clockwork Politics. Fünf Dimensionen politischer Zeit', *Leviathan: Berliner Zeitschrift für Sozialwissenschaft* 43 (Sonderband 30) (2015), p. 28. On politics and periodicity: Scheuermann (*Liberal Democracy*); Robert E. Goodin, 'Keeping Political Time: The Rhythms of Democracy', *International Political Science Review* 19 (1) (1998); Gisela Riescher, *Zeit und Politik: Zur institutionellen Bedeutung von Zeitstrukturen in parlamentarischen und präsidentiellen Regierungssystemen* (Baden-Baden: Nomos, 1994); Dennis Thompson, 'Election Time: Normative Implications of Temporal Properties of the Electoral Process in the United States', *American Political Science Review* 98 (1) (2004); Graeme Orr, *Ritual and Rhythm in Electoral Systems: A Comparative Legal Account* (Farnham: Ashgate, 2015); Elizabeth Cohen, *The Political Value of Time* (Cambridge: Cambridge University Press, 2018), esp. ch. 3.

[23] Claus Offe, 'Temporalstrukturen sozialer Macht', *Leviathan: Berliner Zeitschrift für Sozialwissenschaft* 43 (Sonderband 30) (2015), p. 42. Cf. Friedbert W. Rüb, 'Die Zeit und die Unzeit der Politik', *Leviathan: Berliner Zeitschrift für Sozialwissenschaft* 43 (Sonderband 30) (2015), p. 195: 'die politische Macht [muss] immer durch Fesseln und Bremsen eingezäunt werden, damit sie nicht in Despotismus und Diktatur umschlägt. Ein Mechanismus – wenn auch bei Weitem nicht der einzige – ist Begrenzung von Macht durch zeitliche Limitierung.'

[24] On rhythmic forms as conducive to the maintenance of associations and institutions: Caroline Levine, *Forms: Whole, Rhythm, Hierarchy, Network* (Princeton: Princeton University Press, 2014).

One sees a glimpse of this already in the early-modern origins of periodicity in the temporal circumscription of monarchical authority—prototypically, the 1641 English Triennial Act. While this move to give temporal order to the proceedings of state has more than one significance, at its most basic it was a way to coordinate opposition. By establishing a set of periodic rules within which the King should operate, parliamentarians moved from being a geographically dispersed set of individuals, assembled (or not) on an ad hoc basis, to a potentially coherent collective agent. Something similar is evident in the modern period in the form of the constitution of collective agents such as parties and movements. Patterned intervals of election have been sought out by such agents not just as ways to gain executive power but as ways to coordinate.[25]

An independent political time marked out by institutional rhythms creates reference-points for planning and budgeting, and a sense of when unity is required. Even where the precise duration of alternating elements is variable, rhythmic structures introduce constraints on uncertainty, and shared expectations that a focal-point for action is forthcoming.[26] Such a temporal scheme creates phases of down-time too, e.g. just after an election, when groups may consciously waive the need for unity, allowing opportunities for intra-party debates that would otherwise have particular risks attached. Predictable cycles are also important for the inclusiveness of participation. Where the timing of contestation is patterned, it is less likely to be dominated by the most organized groups, who can afford the permanent mobilization that irregular timing demands. Studies suggest that voter turnout is higher when elections to different chambers are clustered on a single cycle, and lower when these elections are timed irregularly or divided across cycles.[27] Periodicity and participation seem linked.

But perhaps the key challenge for dissenting opinion is to coordinate itself at the level of general principle. Temporally irregular forms of mobilization, be they protests, strikes or referendum campaigns, tend to be broad coalitions of opinion given unity by their relation to particular issues. Lacking a temporal structure independent of events as they occur, they are typically reactive and impromptu, bound up in the particularities of the moment. For the same reason, even where they can adequately contest decisions, they are less able to contest the decisions *not* taken, for which agents must abstract from the passage of events. It is exactly by making the timing of contestation to some degree *independent* of other developments in the life of the polity that institutionalized rhythms help coordinate opinion as competing general outlooks. They facilitate the contestation of power at the level of principle.

[25] Alastair Smith, *Electoral Timing* (Cambridge: Cambridge University Press, 2004), p. 242.

[26] Evitar Zerubavel, *Hidden Rhythms: Schedules and Calendars in Social Life* (Berkeley: University of California Press, 1985), p. 12; cf. Riescher (*Zeit und Politik*), p. 69.

[27] Sarah F. Anzia, *Timing and Turnout: How Off-cycle Elections Favor Organized Groups* (Chicago: University of Chicago Press, 2013), pp. 200–15.

To link back to an observation in our discussion of technocracy, and one we shall return to in the concluding chapter, emergency rule directs attention to the alleged uniqueness of an episode. The logic of exception says that what matters are the details of the moment in its specificity. The critical appraisal of executive decision-making depends, by contrast, on being able to take the opposite approach: stepping back from the details to articulate a position of principle.

A second and related matter is the integration of dissenting opinion across time. Conflicts emerging around one particular issue or decision struggle to challenge the deep structures of power if they cannot be related to other such conflicts, past and expected. It is such interconnections that foster systematic critiques and alternatives. By establishing a temporal structure that extends indefinitely, the institutionalization of periodicity is one mechanism that serves this. It establishes relations between moments of contestation such that they collectively form a series, evoking a whole from the parts. Electoral cycles, debating calendars and the like are all mechanisms establishing links between instances of contestation, allowing each to be located in a larger frame. They integrate the political process such that it is visibly more than a set of isolated clashes.

The state of anticipation, specifically of future opportunities when decision-making and its justification can be contested, is constitutive of how we typically understand political opposition. It is the prospect of opportunities to criticize, influence, or reverse decisions that helps parties develop programmatic alternatives. In established political systems it arguably also motivates 'losers' consent', i.e. the temporary acceptance of undesired decisions.[28] By casting the shadow of impermanence on decision-making, periodicity gives reason to see defeat as something other than capitulation.[29] Likewise, it integrates in the backward-looking direction. It lends political conflict a discernible past, allowing the clashes of the present to be related to previous encounters. Institutional cycles, because they allow temporal abstraction from the current of historical events, provide the conditions for comparison and narrativization.[30]

Legitimation is a final feature one may highlight. An enabling condition of powerful opposition is the recognition of its right to be heard. Sociologists have written of the 'temporal anchoring of normality'.[31] What falls within a periodic structure has its legitimacy undergirded, just as what interrupts such a structure takes the form of a provocation. Periodic institutions make the expression of opposition part of the natural run of things. As John Dewey once wrote in a discussion of rhythm beyond the political sphere, 'There is a wealth of suggestion

[28] Christopher J. Anderson, André Blais, Shaun Bowler, Todd Donovan, and Ola Listhaug, *Losers' Consent: Elections and Democratic Legitimacy* (Oxford: Oxford University Press, 2005).

[29] Offe ('Temporalstrukturen'), p. 42, speaks of the 'integrating consolation-effect of the wait for the "next opportunity"' [author trans.].

[30] Orr (*Ritual and Rhythm*), ch. 3.

[31] Zerubavel (*Hidden Rhythms*), p. 20.

in the phrase "takes place". The change not only comes but it belongs; it has its definite place in a larger whole.'[32] More than just an aid to coordination and integration, rhythms establish that each part *belongs*. They create expectations that are normative, not just predictive. By cultivating the expectation of opposition at regular intervals, periodic institutions foster its acceptance.

Without regular slots set aside for it, dissenting opinion is easily cast as a blot on the efficiency of government. Its legitimacy is liable to be questioned, or acknowledged only insofar as it is does not slow decisions. Irregular forms of contestation, including street protests, referenda, and ad hoc debates, are unable to draw on the normative expectations cultivated by periodicity. The fact they are un-timetabled gives them the character of an interruption. For the very reason that they stand outside the normal passage of institutional time, they present themselves as a distraction from efficient rule, their initiators as the instigators of delay. The delegitimization of critical responses to the emergency measures of the Euro crisis are an example of this pattern in operation. By contrast, by marking out a relatively autonomous temporal sphere, institutions organized periodically protect the practices of opposition from the influence of criteria that put their legitimacy in question. They allow critics to alleviate the time pressures they are subject to by appealing to something standardized.

When institutions are organized on a periodic basis, they operate to a timetable independent of the passage of events. Elections occur because a timeframe has been established for them, either in the shape of a fixed cycle or a range within which it is determined they should occur. The timeframe is broadly knowable in advance, and acts as a common reference-point.[33] Such cycles produce moments of contestation significantly decoupled from external drivers, be they exogenous shocks to which the polity is subject, or endogenous decisions taken by organs of state. Periodic institutions establish moments of contestation abstracted from the temporality of governing itself, and marked off from the current of cultural and socio-economic practices.

Independent of events and problems as they arise, and largely independent of the will of particular agents, periodic institutions establish a distinct temporal order, one that becomes the source of normative expectations. We are dealing not just with factual regularities that exist on paper, as observed by scientific study, but ones that are the object of public appraisal. While institutions in general may display temporal regularities in their functioning—they may have their distinct *Eigenzeit*[34]—what needs stressing is how these become the source of generalized expectations about what will and should happen. This is what the 'temporal

[32] John Dewey, 'The Natural History of Form', in *Art as Experience* (New York: Capricorn, 1958), p. 159.

[33] Offe ('Temporalstrukturen'), p. 44.

[34] Rüb ('Zeit'), p. 196: 'Jede Politik und auch jedes politische Regime hat seine Eigenzeit. Es sind die Wahlzyklen, die Anzahl und die zeitlichen Abstände von Lesungen von Gesetzentwürfen im Parlament

anchoring of normality' entails. Periodicity fosters a temporal order, abstracted from events and agents in their specificity.

From the perspective of technocrats, or those who expect political authorities to conform to markets and other mechanisms of socio-economic power, periodic institutional structures will often seem cumbersome and distracting. One of the longstanding arguments in defence of absolutist monarchy, as a system in which power is *not* contested periodically, was that monarchs can take the long view, unchecked by wider currents of opinion. Such arguments are rehearsed today in the defence of unelected institutions.[35] Executive power responsive to the periodicity of institutions by contrast must reckon with its agenda being periodically interrupted, its problem-solving activities waylaid. But for those wanting to contest decision-making, institutional periodicity has always been a valuable resource. Indeed, critical opinion would seem to depend on there being opportunities for contestation to some degree *irrelevant* to the particularities of rule at a given moment. By abstracting from the passage of events as they occur, and from the timing strategies of particular agents, institutional rhythms enable the contestation of power at the level of general principle.

Contesting Transnational Executive Power

In the transnational sphere, periodicity has traditionally been weaker. One of the problems of EU politics is that it has proceeded with a weaker set of temporal reference-points by which critical agents can organize. Much of what was desirable about the periodicity of politics has been relinquished. Partisanship on a scale coextensive with the scope of decision-making has accordingly been that much harder to achieve. One effect of transnationalization has been to highlight the misalignment of rhythms across countries. National elections in EU states are dispersed in time, while treaty-ratification processes tend likewise to be misaligned. Asynchrony is boundary-reinforcing, encouraging the fragmentation and localization of dissent.[36]

While the European Parliament does offer transnational rhythms, it has tended to be spurned by the more critical party formations, especially of the left, on the reasonable grounds that its institutional powers have long been rather limited. The effect is a weaker set of checks on executive power. Moreover, practices of

bis hin zu anderen Fristen, wie etwa die 3-Wochen-Frist zur Einberufung des Vermittlungsausschusses bei Einsprüchen des Bundesrates gegen vom Bundestag verabschiedete Gesetze.'

[35] Giandomenico Majone, 'Temporal Consistency and Policy Credibility: Why Democracies Need Non-Majoritarian Institutions', *European University Institute Working Paper 96/57* (1996); Tucker (*Unelected Power*).

[36] Zerubavel (*Hidden Rhythms*), p. 67.

emergency rule aggravate the situation. Exceptional arrangements such as the Troika, and policy regimes of uncertain duration led by technocratic institutions, further limit the formation of institutional periodicity.

The conclusion one may draw is twofold. On the one hand, *non-periodic* mechanisms become increasingly important for the contestation of power, at least on an interim basis. Temporally irregular forms of mobilization are what transnational partisans must fall back on, and they are not without their own important qualities. On the other hand, the question is whether the periodicity of transnational politics can be fostered. Some emerging partisan formations are taking steps to this effect, and in doing so may build the conditions for stronger opposition in the longer term.

To begin with, consider how, when dissent is raised in the contemporary European context, it often borrows elements of a different temporal scheme from the periodic. Both the right-wing populism and the leftist politics of disobedience examined in previous chapters have involved the appeal to referenda. Such devices tend to be premised on the uniqueness of the moment. Think again of the referendum called by the Greek government in July 2015 on the bail-out measures demanded by the country's creditors, with its language of the 'historic opportunity'[37]—of the moment that comes only once, of actions now or never. The emphasis is on an opportunity to contest that is not to be repeated, at least not to a foreseeable timetable.

Such forms of opposition to EU decisions embody a distinct temporal structure. If they have a common denominator, one might call it the principle of the *threshold*. Rather than being activated on a periodic basis, contestation on this model occurs whenever a certain parameter is crossed. The threshold may concern the magnitude of the decisions contested (the sense for instance that they are of constitutional significance), or the degree of critical public opinion they solicit, or the novelty of the issues they raise. The moment of heightened contestation arrives *whenever* such thresholds are crossed (or never).

Proposals for institutional redesign often point in the direction of further embedding the threshold model. Efforts to enhance direct democracy typically take this form. Obligatory referenda, triggered whenever decisions of a certain constitutional magnitude are advanced, or citizens' initiatives and counter-referenda triggered whenever a certain level of public support has been expressed, are mechanisms of a threshold kind.[38] Although lacking in binding force, the EU's European Citizens' Initiative (ECI) is one example at the transnational level of a mechanism apparently geared to the threshold principle, as something that kicks

[37] Tsipras, 1 July 2015: https://archive.change4all.eu/change-in-words/detail/prime-minister-alexis-tsipras-message.html/.

[38] David Altman, *Direct Democracy Worldwide* (Cambridge: Cambridge University Press, 2011). In the EU: Lacey (*Centripetal Democracy*); Francis Cheneval and Mónica Ferrín, 'Referendums in the European Union: Defective by Birth?', *Journal of Common Market Studies* 56 (5) (2018), pp. 1178–94.

into gear when a certain number of signatures is achieved. In such forms, the contestation of decision-making is pursued without a temporal frame based on repetition.

Mobilization on the threshold model is in some ways well suited to contesting power in an age of emergency rule. Being independent of the electoral cycle and other institutional rhythms, it can contest emergency decisions in real time. Some such modes can be organized cross-nationally, exactly because they are separate from the non-synchronized cycles of institutions. Extra-parliamentary protests and strikes can be arranged simultaneously across borders more easily than electoral mobilizations. Moreover, given the timing of elections will not reliably coincide with moments of heightened executive discretion—indeed, given the latter may sometimes be timed to avoid this—it is hard to see how emergency rule can be constrained without the involvement of these one-off mechanisms of contestation. By cultivating a more general critical awareness, such actions also have the potential to spill over into something enduring.

But can opposition on the threshold model substitute for opposition on the periodic one? Even acknowledging the unexplored potential of direct-democracy mechanisms, this seems unlikely. Critical opinion under such conditions, to the extent it is able to coordinate, will tend to do so around demands in their specificity. The focus is on the situation at hand, on the impulse that triggered the critical response. Such forms of dissent are liable to recede as quickly as they emerge. Lacking an expected follow-up, they struggle to embed themselves in a larger narrative frame. It becomes harder to reveal the stakes of individual decisions by locating them in historical context, and to mobilize supporters by cultivating anticipation of the next encounter. Executive agents have little reason then to factor into their decisions the expectation of substantial opposition.

To what extent then might one expect a sharper *periodic* form transnationally? One possibility is its gradual emergence without any particular set of agents willing it. It has been suggested that early signs are visible in the EU, as the cycles of technocratic and political institutions—notably the Commission and Parliament—start to synchronize.[39] If such a trend is real, the misalignment of national electoral cycles and processes might start to matter less. Certainly, there is a difference between rhythms on paper and those that have public resonance. The emergence of the latter on a transnational scale remains a nascent development at best, and weakly aligned with the enduring temporal structures of states. But their emergence would be a positive development.

The question is whether such trends could be extended further by concerted efforts to capitalize on existing institutional resources and what rhythms they afford. Here one can imagine several approaches, all the while staying close to

[39] Klaus Goetz, 'Synchronisation demokratischen Regierens in der Europäischen Union', *Leviathan: Berliner Zeitschrift für Sozialwissenschaft* 43 (Sonderband 30) (2015), pp. 132–51.

existing constitutional arrangements. One of the more obvious, if difficult to achieve, concerns the harmonization of national electoral cycles. Given the enduring centrality of state governments to EU power, holding national general elections on the same day would be a significant move in reinforcing cross-border periodicity, and with it a more cohesive and transnational contestation of power.[40] While the harmonization of all member-state cycles is clearly a remote possibility, absent the consent of those who have reason to resist it, the synchrony of a few is something like-minded partisans organizing across borders could pursue. Depending on the electoral system and the distribution of power within it, opportunities might arise through pressing no-confidence votes on a weak government or forcing the breakdown of coalition negotiations. Though largely agent-independent, such cycles may be influenced at the margins.

A more promising way to recuperate periodicity at a transnational level involves making use of the *non*-synchronization of national electoral cycles. Misalignment has its own potential value. As a point of departure, consider the fact that somewhere in the EU, in one member-state or another, there is always an election around the corner, and therefore a potential constraint on the European Council. Approximately six or seven times per year, the composition of a Council member is contested in a national election.[41] So long as national elections are contested only by national parties, i.e. those organized at this level and advancing programmes focused on domestic policy, it makes little sense to see these elections as aggregating to form cross-border rhythms. But a transnational partisan movement that contested multiple national elections on a rolling basis would invite a major shift in perspective. It would then become possible to speak of periodicity on a wider scale, each cycle of European politics lasting the interval of one national election to the next, and the foundation on which a more lasting influence on EU-level decision-making could take shape.[42] Amongst the beneficiaries would be transnational partisans themselves, able to use these recurring moments of contestation to coordinate, integrate, and legitimize their activities.

One tends to associate transnational partisanship either with supranational chambers or a certain kind of intergovernmental cooperation, but a transnational movement or alliance pursuing a networked campaign across national elections, articulating a transnational perspective in each, would be an intriguing variation on the theme. Beyond the use of a common name across different elections, such

[40] For a proposal to this effect, though intended to facilitate more speedy executive decisions: Giorgio Basevi, 23 January 2013 (https://voxeu.org/article/reducing-frequency-electoral-cycles-eu-proposal-synchronising-national-and-european-elections). On the problems of misaligned electoral cycles: Lacey (*Centripetal Democracy*), p. 118.

[41] I.e. in the general elections of c. 27 member-states.

[42] National elections in the EU are of course not quite evenly spaced, and different countries have different electoral systems. Also, the elections of some member-states are always likely to have greater ramifications than others. Notwithstanding these caveats, something like the benefits of periodicity as described in the previous section would seem to be offered on this model.

formations would seek to connect local campaigns to a common programme of reform. Extending party membership across national borders so as to cultivate a multinational membership group would be one way to underpin this organizationally. Such ideas have been aired in the debates surrounding DiEM25, albeit mainly in the abstract. At the time of writing there are also smaller formations configured on these lines which aim to build transnational partisan associations that compete in national elections.[43] Although research is limited, evidence suggests a constituency may exist in national electorates for transnational mobilizations of this kind.[44]

There is a historical precedent of a sort. It comes from a quite different context and era, namely the Arab Socialist Baath Party. Founded as the electoral wing of a social movement in the late 1940s, the party developed branches in Syria, Lebanon, Iraq, and Jordan, with the aim of standing in national elections. At least in its early years of existence, this was an ideologically-cohesive partisan association, centred on opposing colonial power, pursuing pan-Arab political unification, and developing socialist policies to alleviate poverty.[45] Its agenda was to contest the coordinated exercise of repressive executive power across a wide territorial space—somewhat analogous to the case at hand. This particular experiment in transnational partisanship ended badly, in part due to challenges that would face any such initiative, including how to maintain programmatic unity at critical moments and how to contend with unequal power relations between states. But if such a party were able to consolidate itself as a genuine unity, and to do so without compromising its popular appeal, it would potentially become an important check on transnational executive power and a foundation for efforts to seek direct control.

The Europeanization of national elections on this model would in some ways mark an *improvement* on conventional party democracy, in particular as regards the constraint of executive discretion. Long inter-electoral spaces of four years and more are often cited as one of the weaknesses of electoral politics in the state setting.[46] In an age of emergency politics, where decision-making is accelerated and socio-economic conditions are volatile, those inter-electoral spaces look longer than ever. A lot happens from one election to the next. By contrast, the abundance of national elections in the EU means the interval from one to the next is much shorter. When bridged by transnational partisans, they add up to a more

[43] See e.g. Volt (https://www.volteuropa.org), esp. the movement's 'Amsterdam Declaration', 18 October 2018.

[44] Lucy Kinski, 'Whom to Represent? National Parliamentary Representation during the Eurozone Crisis', *Journal of European Public Policy* 25 (3) (2018), pp. 346–68.

[45] John F. Devlin, 'The Baath Party: Rise and Metamorphosis', *The American Historical Review* 96 (5) (1991), pp. 1396–1407.

[46] Altman (*Direct Democracy*), p. 1: 'the time elapsed between elections may be agonizingly long for citizens whose preferences are systematically unheard, and these interelection spaces constitute the weakest link of current democracies'.

densely packed series. With minimal change to existing institutional structures, one could thus envisage an accelerated pattern of contestation with more regular expressions of opposition.

If these are long-term projects, there are also more immediate ways in which mechanisms of cross-border periodicity can be cultivated. Existing institutions such as the ECI are in principle susceptible to being re-appropriated as mechanisms of periodic rather than threshold-based contestation. Parties are well-placed to initiate ECI actions, and cooperating transnationally could bring proposals on a recurrent, e.g. annual or bi-annual, basis.[47] Periodicity could in this way be enhanced by reinterpreting existing institutional structures. In addition, mechanisms of partisanship itself can be configured on the same temporal structure. Transnational conferences are a way to establish recurring moments of visibility and engagement, a mechanism used by the Internationals of the past. Synchronizing and regularizing the timing of national party conferences, or harmonizing certain policy statements, would enhance the attention they receive. Such rhythms of initiative and assembly are ways to consolidate unity amongst the like-minded and heighten their visibility to a wider public.

Clearly the European Parliament remains one of the more promising sites for the consolidation of periodicity on a transnational scale, whatever its current limitations. Its five-year cycles are clear and fixed, and coextensive with the EU as a whole. If one allows oneself to imagine major redesign, strengthening the Parliament such that its rhythms of election and debate can better impose themselves on decision-making would be conducive. Increasing its powers vis-à-vis the Commission, even embedding the latter within it, has much to commend it in principle. So too does strengthening the Parliament's power to initiate legislation, elevating it relative to the Council.

But even in its current form, as an institution easily side-stepped by executive power and whose law-making powers are patchy, the Parliament offers a calendar around which groups can organize. By addressing themselves to the Parliament's cycles, partisans may undergird the transnational character of their activities and their long-term horizon. While the established party groupings of the Parliament have generally offered little in this sense, being weakly defined ideologically and rather detached from mass publics, far-Right parties have used this arena to build their identity and influence. The 2019 EP elections reaffirmed how the campaigning process offers more than just the chance to win seats. In DiEM25 moreover, as well as amongst Greens, one saw how such an approach might be replicated from the left. Rather than as just a means to win office, activists conceived EP elections as a way to mobilize critical opinion and consolidate their organizational form, mapping out the EU's political future and

[47] Lacey (*Centripetal Democracy*), p. 234.

their own trajectory within this.[48] By offering dates independent of the flow of events in their specificity, the Parliament's periodic form offers such groups a way to integrate their activities across time and seek to build an enduring presence.

All such opportunities depend for their exercise on partisans committed to a transnational view. They depend on a capacity to develop shared substantive programmes of diagnosis and advocacy, linking a transnational agenda to local causes. With its entrenched structures of political and socio-economic power, the EU is an inhospitable terrain, and projects of transnational partisanship will no doubt tend to falter. But arguably it is only now, in the wake of the EU's crisis politics of the 2010s, that they are starting to be pursued in renewed earnest. The project of constraining transnational executive power is at an early stage, and in the creative engagement with existing mechanisms one sees the contours of a plausible new template. Organizing transnationally to contest European arenas—notably the European Parliament—here combines with something potentially more significant: the contestation of national elections on a transnational programme. As established and regularly recurring moments of contestation, such elections are arguably the best platform on which to contest executive power in the EU, both to oppose it and in the longer run to enter and transform it.

Conclusion

The politics of emergency in the transnational setting expresses the detachment of executive power from a range of constraints. One of these is the political party, whose widely observed hollowing in the second half of the twentieth century corresponds to many of the tendencies examined in this book. Warding off recourse to government by emergency would seem to depend on strengthened ties of partisanship, both as these may constrain representatives directly, and as they may influence executive power from outside.

It may be that without a deeper process of constitutional reform or refounding, mobilized responses of this kind will make limited progress. Indeed, without far-reaching changes to underlying socio-economic structures, even constitutional responses may come up short. But any such schemes of redesign have to be conceived in conjunction with the agents that could lead them and preserve their achievements. Even a radically improved political and socio-economic order would need to be subsequently maintained. And should the EU ultimately

[48] See the commentary by one of DiEM25's founders, Lorenzo Marsili, 14 November 2017 (http://politicalcritique.org/world/eu/2017/can-diem25-become-the-first-transnational-party/): 'the European elections are an excuse, a device. There is little interest in sending a dozen people to a heavily restrictive parliament. The aim is rather to "hack" the elections, to transform them into a continental political campaign: a kind of "Sanders moment" for Europe.'

disintegrate, there is every chance this too will be the occasion for an executive power grab in the name of a transnational emergency. In one permutation or another, contending with practices of emergency rule implies a role for collectives that resist the concentration of power in a small group of leaders. It implies enduring agents of political mobilization—agents of partisanship.

In reflecting on the conditions that can support this, I have focused on the temporal dimension. Associations come alive by virtue of their rhythms: this has been true of contestation in the state context, where the periodicity of institutions has been an important underpinning. Opposition needs reference-points for coordination, and resources to link present conflicts with those of past and future. The periodic form offers valuable resources.

I have emphasized these temporal aspects also because they intersect with some of the more problematic legacies of emergency rule. With its urgency of word and deed, emergency rule signifies the subordination of political decision-making to the temporality of the socio-economic sphere. One of the contributions of institutions of periodicity in the state context has been to mark out the autonomy of political time. By establishing time-structures independent of the passage of events, such institutions offer a framework for the coordination of opinion at the level of general principle, abstracted from the specificities of a given context. By making time for opposition, they also undergird its legitimacy.

In the transnational sphere, such a temporal order has been weaker. This is one reason why emergency rule goes weakly opposed, for critical citizens lack an important resource. To the extent that dissent is still mustered, it tends to take non-periodic forms, centred on the temporality of the threshold. Structuring partisan power depends on the emergence of clearer political rhythms at a transnational level. By generating some such rhythms of their own, as well as making the most of those already in play, those critical of executive discretion under the sign of emergency can take some steps towards its political constraint.

9
Beyond a Politics of Singularities

The politics of emergency is widespread in today's world, in spheres remote from its conventional association with war. Logics of security and exception have come to pervade the governing of economy and society more broadly. In the workings of the EU and its member states, one sees how they entwine with the dynamics of the transnational sphere. Structural features of this setting ease the passage of reforms that maintain and reinforce the basics of the existing order, while the same shifts find their rationale in problems of cross-border dimensions. EU politics offers a rich perspective on this latest iteration of a long-standing governing template, as well as some insight into the responses it may draw.

My aim has been to step back from some of the immediate implications of emergency politics to consider its larger contexts and legacies. One way to draw together the strands of the book is by highlighting a persistent theme of the period in question, centred on a politics of the particular. Emergency rule is conducted and narrated as the encounter with unfamiliar situations that demand to be handled on their own terms. It is about doing things differently because the situation at hand is different. At least in terms of its own rationale, but also in view of the creations it gives rise to, emergency rule is geared to the singularity of a certain moment.

For those who seek the positive in crisis politics, exactly this idea of a polity evolving with unfamiliar circumstances may be part of its general appeal. There is an experimentalist view of politics in which improvised responses to untried conditions are the essence of what governing is about. Such situations are those in which creativity must be exercised and from which learning is possible. As we have seen, parallel to the argument of this book is one that sees crisis politics as how the EU has positively matured over time. Such ideas of integration-through-crisis are appealing because they suggest, not just that extreme circumstances can be overcome, but that they can be politically enriching too. As we have also seen, there is an authority claim well suited to when decision-making is geared to the specificities of an irregular situation. The notion of *crisis expertise*, in which distinct forms of know-how must be exercised because the conventional formulas do not apply, speaks to this theme of creativity in exceptional circumstances. Whatever the rights and wrongs of particular emergency measures, some would urge us to look for the bigger picture and see a model of dynamic evolution that deserves appreciation.

Politics of Last Resort: Governing by Emergency in the European Union. Jonathan White, Oxford University Press (2020).
DOI: 10.1093/oso/9780198791720.001.0001

Yet what also becomes clear from our analysis is how some of the major challenges of emergency politics are connected to the same insistence on particularities. From Monnet's 'provisional European government' to the Eurozone Troika, major reconfigurations of executive power and policy are pursued in the name of the distinctive pressures of the moment. Democratic norms tend to be one of the first casualties, as procedural constraints are evaded to advance measures that otherwise might meet greater resistance. Although much of the scholarship on emergency rule takes a liberal approach, focusing on the harms done to individual rights, it is clear that executive usurpations of power tend to further remove decision-making from popular control. The challenge to responsive and accountable rule is arguably deep-seated and lasting.

One thing the politics of exceptionalism makes visible is democracy's uneasy relation to the unique. Embedded in references to exceptional conditions is the idea of a situation that resists the usual comparisons and standards. Rather than a conception of history centred on continuities, implied is an experience of time marked by episodes discrete and unalike. Insofar as political theorists have reflected on the significance of the unique, emphasis has tended to fall on the enabling aspects. Revolutions and constitutional foundations have been analysed as moments that, precisely in their singularity, herald new political and moral horizons. But emergency rule brings to the fore a more troubling aspect. Many of the challenges it poses are linked in some way to its characteristic emphasis on the *sui generis*.

Situations experienced as unique, and framed as such by powerful actors, become harder to locate in normative frames. Compartmentalized in time, and symbolically marked off from an anterior and posterior normal, such episodes resist contestation at the level of principle, for which a sense of equivalence between events is crucial. Unconventional governing responses in turn further isolate the period in question. The obscuring of long-term causal factors, and the tendency to ward off radical change, are outcomes related to this. Such an outlook undergirds an idea of politics as a reactive exercise, based on responding to problems as they arise rather than the setting of general priorities. By detaching aspects of the present from a larger perspective, exceptionalism contributes to a political culture shaped around events in their specificity.

Part of the logic of emergency more generally, an emphasis on the unique has special resonance in the EU context. As an association of relatively recent construction, the reference-points that might foster historical analogies are sparser. Comparisons in space are hindered by the rarity of such unions and the long-standing tendency to see the EU as an association *sui generis*. Moreover, its fragile constitutional arrangements mean the distinctiveness of the moment is all too easily reinforced by novel executive formations like the Troika that lack precedent in earlier practice. The continuities on which democratic evaluation depends are thereby disrupted and pre-empted. EU politics comes to be viewed as an aggregation of crises and crisis measures—of economics, migration, populism, and more.

To be sure, there are features of today's world that are importantly novel and that deserve to be appreciated as such. Economic developments of the last decade and more bring home the deep-rooted problems of the socio-economic order in a peculiarly clear form, and the effects of these and the responses they have occasioned are far-reaching. But the challenge is to acknowledge their distinctiveness without embracing a debilitating view that detaches them from context. Those features of the socio-economic world that are enduring are harder to recognize as such against the backdrop of rapidly changing institutions cast as responses to fast-changing circumstances. This is the deeper sense in which emergency practices marginalize larger questions of principle and the contestation of power.

The responses to emergency rule discussed in the previous chapter, centred on partisanship and the rhythms of opposition, are not just ways to constrain executive discretion, but to envisage a more general shift in political culture beyond a politics of singularities. When the present is approached as a time of unique challenges, it is severed from evaluation by appeal to the past and becomes a weaker foundation for future-oriented projects. The basis for long-term planning and progressive change depends ultimately on the thought that the concerns of the present, rather than exceptional and a symptom of uncertainty, are continuous with those of a world to come. A constructive outlook on the future is one of the less appreciated features of democracy that emergency rule undermines, and good reason to resist the temptation to see crisis politics as something productive.

Democracy and the Unique

Exceptionalism entails releasing a particular case from the application of a more general norm.[1] We have seen there are at least two dimensions, to do with both the circumstances at hand and the way they are governed. On the side of decision-making, one sees it in the form of 'exceptional measures' that depart from pre-existing procedural norms. The suspension or disruption of conventional methods and the improvisation of new ones are characteristic features. On the other hand, one sees it in claims made about the socio-economic conditions, where 'exceptional circumstances' are evoked as the warrant for measures taken. These two dimensions are often conflated, but it is important to keep them separate. We have seen that while the first is constitutive of emergency rule, and must therefore be

[1] Agamben (*State of Exception*), p. 25. Cf. Davies (*Limits of Neoliberalism*), pp. 174–5.

assumed as a precondition for this analytical frame, the second is the contestable basis on which it is rationalized. The agents of emergency rule claim exceptional conditions as their warrant; the true nature of those conditions is a political question.

When circumstances are hailed as exceptional, in principle they can still be approached as part of a recurring series. However rare, unfamiliar and even unforeseeable they may be, extreme circumstances need not be viewed as unique.[2] In the Roman republic, emergency rule was typically used as a way to fight wars, and although such conditions were understood to be exceptional, they formed an identifiable pattern nonetheless. Wars had some predictable features, including a duration shaped by the military campaigning season. They tended to resemble each other in some measure, and because the underlying crisis was understood to be structured, emergency rule itself could be the stuff of conventions and shared expectations. The dictator traditionally had a term of office limited to around six months. Citizens caught up in the practices of emergency rule had some idea then of what they were contending with.[3]

In the transnational context we have been considering, one sees traces of the same idea of circumstances that are exceptional but not wholly unprecedented. It is evident whenever one sees references to financial crashes as events that recur. Cyclical accounts are a mainstay of contemporary liberal economics, endorsed by both scholars and policy-makers, and have been used to suggest lessons that can be drawn from history.[4] Likewise, notions of recurrence are implied whenever comparisons are made to specific events in the past, e.g. to the Great Depression. Such historical reference-points were no doubt on the minds of Europe's policy-makers in the 2010s, arguably encouraging a preference for certain policies over others.[5] On such views, exceptional times are abnormal times, comparable only with the most extreme of past experiences, but not without analogues entirely.[6]

But the transnational context also seems to invite a stronger view, whereby the circumstances at hand are cast as unprecedented in scale or kind. They resist meaningful comparison with other cases. This view was central to the rhetoric and

[2] Eger, Oeter, and Voigt (*International Law*) distinguish 'states of emergency', which are potentially foreseeable and regularized, and 'states of exception' which demand improvised responses.

[3] Lincott (*Roman Republic*); though for the argument that the six-month rule was emphasized more by later interpreters than contemporaries: de Wilde (*Dictator's Trust*).

[4] As influential economists approaching the events of the 2010s as the latest in a series: Carmen Reinhart and Kenneth Rogoff, *This Time is Different: Eight Centuries of Financial Folly* (Princeton: Princeton University Press, 2009). Whether they draw the right lessons from past events is another matter.

[5] E.g. Barry Eichengreen, *Hall of Mirrors: The Great Depression, The Great Recession, and the Uses-and Misuses-of History* (Oxford: Oxford University Press, 2015), esp. pp. 382ff.

[6] We have noted the tendency of policy-makers to cast problems as quasi-natural forces—economic storms, migration waves. This might seem consonant with the idea they recur, insofar as the natural world is marked by rhythms and cycles. But whether, in the age of climate change, nature still stands as a figure of predictability is questionable. Arguably such metaphors have come to evoke exponential change and ever more extreme situations.

practice of the period examined, and has partly to do with the circumstances at hand. Emergency rule in the transnational context is targeted at challenges of a socio-economic kind, classical military questions generally being beyond the scope of transnational institutions. In the socio-economic field, extreme circumstances are harder to define: they lack unmistakeable boundaries in time, and lack a clearly defined spatial delimitation.[7] There is nothing self-evident about the crises invoked: whatever violence they entail is for the most part invisible. Their identification depends on the mediation of authoritative interpreters, who may have reason to emphasize the singularity of the moment so as to press for executive action. In such contexts, if exceptional circumstances are to be cited, often it will be as a unique configuration rather than the latest instance of a recurring kind.

The novelty of the moment tends to be underscored also because of the problems of authority that transnational politics contends with. For executive agents of the EU realm, whose authority to act is uncertain at the best of times, departing from procedural norms to improvise new arrangements demands the circumstances be cast as especially unusual and challenging. If they were but the latest instance of a familiar kind, the presumed licence to deviate from norms would be weaker. As we have seen in the crisis discourse of monetary and migration politics, the emphasis is accordingly on the singularity of the moment—the unprecedented nature and scale of the problems.[8] Securitizing agents highlight the most dramatic and fast-moving features to signal the distinctiveness of the situation. Emergency rule in this context tends to be a politics of the declaredly unique.[9]

The sense of uniqueness is reinforced by the nature of the political response. The particularity of the case is proclaimed, but also performed by the use of unconventional measures. Existing frameworks are selectively suspended or ditched, in line with the thought that the world they belong to is different from the situation at hand. We noted in Chapter 1 the distinctly ad hoc character of such arrangements as the Troika and ESM. These were structures discontinuous with previous arrangements, refashioning the identity of executive power. Such

[7] Whether contemporary military conflicts are not also increasingly amorphous is something we may bracket. For Manin, terrorism is a contemporary challenge unsuited to emergency rule precisely because of its protracted nature (Bernard Manin, 'The Emergency Paradigm and the New Terrorism: What If the End of Terrorism Was Not in Sight?', in Sandrine Baume and Biancamaria Fontana (eds), *Les usages de la séparation des pouvoirs* (Paris: Michel Houdiard, 2008)).

[8] Even when authorities declare the need for exceptional measures towards phenomena that recur—e.g. migration—they generally suggest the challenge is qualitatively different this time, that the circumstances are unprecedented. See e.g. European Commission (2016), 'Proposal for a Council Implementing Decision . . .'.

[9] For precision note that, while *unprecedented* suggests circumstances without precursors, it retains the possibility they may recur (and hence of a learning process); *unique* implies circumstances that will not.

improvisations, with a beginning and an expected conclusion, carve out a distinct episode in the passage of institutional time.

'A common approach has not developed at European-Council level for managing crises. As each crisis has different causes, is based on a different problem and is made up of specific, often unique, circumstances, each is dealt with differently.' Such was the conclusion of a 2016 European-Parliamentary report, neatly encapsulating the logic of EU crisis management.[10] At some level of analysis, everything is indeed unique. Every situation has its specificities, to be assessed against a moving historical backdrop. Connecting circumstances across time relies on an exercise of judgement and a willingness to find continuities. To draw links through the application of common rules, norms, and principles involves a concerted act of generalization.[11] And there are cases where the discontinuities may be as striking as the continuities. But exceptionalism entails radicalizing this observation, emphasizing the distinctive features of a situation and restricting comparisons, if made at all, to historical episodes like the Great Depression that are prototypes of the extreme and unparalleled.

A theme in certain strands of political philosophy has been that singularity can be a productive phenomenon. Cases that resist the application of an established norm can have a positive, norm-*generating* quality. Some stress that situations and decisions initially experienced as singular can retrospectively acquire the 'force of the example', generating new standards by which they and analogous cases may be judged.[12] Even if discontinuous with what precedes them, they can stand at the beginning of a new series of equivalencies. The modern idea of revolution pioneered in eighteenth-century France and America, understood as marking rupture rather than return and offering a template of progress for others to follow, is a paradigmatic instance of this.[13] In this sense, the singular can be enabling.

Emergency measures can display something of this generative quality. Although priority tends to fall on solving immediate problems (and the setting of precedents may be rejected), a more constructive dimension can emerge unintended. Although shoring up existing power structures is generally the first aim, deeper transformations may follow. Our discussion of principled disobedience, including its demos-building possibilities, indicated some of this potential in the politics of emergency 'from below', just as it may be present in emergency rule

[10] 'The European Council and Crisis Management', European Parliamentary Research Service, PE 573.283, February 2016, p. 26.

[11] Alessandro Ferrara, *The Force of the Example* (New York: Columbia University Press, 2008), p. 52; Sunstein (*Legal Reasoning*), pp. 63ff.

[12] This scholarship draws on Kant's theorization of aesthetic judgement and Arendt's engagement. See Ferrara (*Force of the Example*), esp. p. 3; cf. María Pía Lara, 'Reflective Judgment as World Disclosure', *Philosophy and Social Criticism* 34 (1–2) (2008), p. 89. For a critical take on 'contemporary political theory's identifications of action with event, and politics with singularity': Honig (*Emergency Politics*), p. xviii.

[13] Hannah Arendt, *On Revolution* (London: Penguin, 1990), pp. 27ff.

'from above'. But to the extent that there are such normative dividends, these are inevitably deferred to the long-term, and somewhat unpredictable in makeup. What needs to be underlined are the difficulties that arise when decision-makers govern as though the circumstances are unique. Discretion exercised on these terms poses some distinctive challenges, above and beyond its substantive shortcomings.

Approaching circumstances as though they are unique is one way for powerful agents to avoid justifying their decisions. To assert the particularity of the situation is to question the possibility of applying wider criteria.[14] When decisions are rationalized as addressing unprecedented conditions, there is typically no claim that they are desirable against context-independent principles. The suggestion is simply that they are necessary for the situation at hand: they are needed to maintain something's existence.[15] The practices of public justification one associates with even the most minimal idea of democracy are hard to square with this. By detaching the specific case from a larger context, the notion that larger value questions might be at stake beyond the problem-solving demands of the moment falls away.

From populations, this outlook demands an attitude of trust. Emphasis on the singularity of the moment interrupts the capacity to connect decision-making to values and principles widely endorsed and to contest initiatives on these terms. It is hard to criticize measures for which no general claim to desirability is made. Decisions become grounded in the knowledge of details and particularities that only insiders are well placed to claim. Control of the agenda is more tightly maintained in the hands of the few. Though governing agents may still present their decisions as grounded in expertise, the claim becomes hard to evaluate. Denying the applicability of general criteria points to an increasingly unaccountable form of rule.

To govern as though the circumstances are unique is challenging in more lasting ways too. It puts in question the pursuit of consistency in decision-making across time. As Scheuerman observes, a heightened sense amongst policy-makers of fluidity and uncertainty corresponds to the weakening of 'a commitment to general, clear, relatively stable, prospective general rules' and moves away from legal stability or evolution towards serial ruptures and ad hoc fixes.[16] If the present is held to be unique, the legal inheritance from the past is less useable, and there is less reason to draft legislation that can apply to future cases. Though policy regimes may be firmly embedded, scope is left for the exercise of discretion at

[14] Cf. Sunstein (*Legal Reasoning*), pp. 136ff. He refers to 'full particularity' as when 'outcomes are not merely incompletely theorised; they are not theorised at all. They are based on a special form of casuistry—judgement unaccompanied by reasons uniting or distinguishing cases.'

[15] One reason agents may avoid justification in these situations is because they are concerned these will later be taken as precedents.

[16] Scheuerman (*Liberal Democracy*), p. xvii; cf. Opitz and Tellmann ('Future Emergencies').

key moments so that their application can be selective. Again, this weakens the capacity to hold executives to account.

As long as decisions are in the hands of elected officials and those they have delegated responsibility to, some will argue that such concerns can be overlooked. To the extent that executive power in the EU has the representatives of national governments at its core, and improvised arrangements such as the Troika are formally responsive to the Council, executive discretion under the sign of emergency arguably retains a democratic pedigree.[17] But while the electoral authorization of officials is evidently important, its significance is diminished when the decisions they take resist connection with the reasons for which they were elected. This is what the logic of exception entails. In the name of crisis management, decisions of long-term significance are adopted, yet—because the challenges to which they respond are held to be sudden and fast-changing—generally with weak relation to the governing programmes for which executives were elected or the mandates for which they were constituted. As these agents readily argue, the problems they face were unforeseen. Forswearing fidelity to existing commitments, they exercise an especially elastic discretion.

Up to a point, political improvisations in the name of unfamiliar circumstances may be a fair response to a changing reality. But when emphasis falls on the unprecedented and the unique, inevitably what are obscured are the constants and connections. By casting off such continuities, emergency thinking prioritizes the pursuit of order over considerations of what kind of society will be produced along the way. By directing attention to an expected point of resolution when normality is restored, it prioritizes the search for fixes rather than deeper transformations. Progressive politics has relied on an idea of continuity between the problems of the present, past, and future.[18] Sacrificing it comes at a significant price. Enduring normative concerns, and relatively unchanging relations of power, become that much harder to identify, along with the sustained political efforts that may be needed to engage them.

Domination by Exception

Practical expressions of the challenges of exceptionalism are all too visible in Europe's crisis politics of the 2010s, and here it will be enough to restate some of the book's themes. One of the hallmarks of emergency rule, and a corollary of defining circumstances as unprecedented, are measures of uncertain duration.

[17] On the claims of EU institutions themselves, notably the Commission and ECB, that their actions of crisis management were consistent with democratic accountability: Rittberger ('Integration without Representation').

[18] Robert Nisbet, *History of the Idea of Progress* (London: Routledge, 1994), p. 91.

When a situation is held to be unique, implied is that the timespan of its challenges is unforeseeable. Logically, the duration cannot be confidently gauged, since there is an absence of relevant information to go by. Measures of response are accordingly presented as 'temporary', a flexible concept that can denote days, weeks, months, or years. Whether executive agents are genuinely uncertain is not the point. Even improvised forms of crisis management may be undertaken with a provisional sense of the intended timescale. Important rather is that by adopting the emergency framing they have a rationale for being evasive. They can justify—to opponents, to a wider public, and indeed to themselves—ambiguity regarding the duration of the initiatives they undertake.[19]

Varoufakis has written of the challenges posed by the 'permanent temporariness' of policy-making in the Euro crisis, and how ambiguities of temporal duration were used by Troika and Council officials as a means to ward off opposition.[20] The temporal uncertainty embedded in the concept of the exceptional makes it hard to distinguish that which is to endure from that which is merely an interim measure, and hence for opponents to direct their actions productively.[21] Often it will be unclear whether to contest a policy or merely tolerate it, since its discontinuation in the near future may or may not be on the cards. Such ambiguities in turn make it harder to raise a critical public response, even where long-term implications are suspected.

An uncertain temporal horizon is just one of the ways in which exceptionalism spells problems for contestation. Another under-remarked aspect is how emergency rule is often a time of linguistic innovation. One sees new terms introduced and popularized to describe the specificity of the challenges faced and the responses appropriate. In addition to the 'Troika' itself, consider such phrases as 'systemic risk' and 'structural deficit', which became central to diagnoses of the Euro crisis, or the 'macroeconomic imbalance procedure' as part of its management.[22] Recall the concept of 'the spread', that rose to prominence in public debate as a symbol of increasing market uncertainty and the need for certain reforms. To use new vocabulary is to give descriptions of the world that are detached from existing conceptual and normative frameworks and require time to be absorbed.

Unconventional institutional arrangements trade on and reinforce the singularity of the moment. When political responses are improvised in *sui generis*

[19] On such dynamics at work in the emergency politics of the American Civil War: Sabine Mischner, 'The Temporalities of Exception. The Long Shadow of the American Civil War', *Zeitschrift für* Politikwissenschaft 28 (4) (2018), pp. 439-451.

[20] Varoufakis (*Adults in the Room*), p. 304.

[21] Recall the ECB's stated policy of buying the bonds of indebted Eurozone member states (the Outright Monetary Transactions (OMT) scheme): its expected duration was unclear when announced in August 2012.

[22] On the 'structural deficit': Martin Wolf: https://www.ft.com/content/f54332d2-66dc-11e1-9e53-00144feabdc0, emphasizing the term as new, imprecise, and hard to measure.

formats, on the model of the Troika, the Eurogroup, or Memoranda of Understanding (MoUs), there is no existing template for a critical public to apply. The use of unconventional structures marks off the period in question as a world unto itself. Perpetual shape-shifting is a way by which executive power can slip the constraints to which it might otherwise be held. That such formats are often task-specific, tied to the achievement of certain goals, further embeds the singularity of the period. Rather than the basis of generalizable rules, they are practical solutions to particular problems.[23] They detach the period in question from the flow of political time and cast it as a distinct episode, one that can be aggregated but not integrated with others.

When exceptional measures for exceptional times are pursued in a conventional state setting, several factors work to counter this. The polity is generally of longer standing than the EU, affording an historical view that can be used to challenge notions that the circumstances at hand are unprecedented.[24] A more settled institutional structure means that the unconventional measures adopted are at least adopted by familiar executive agents. These factors are weaker in a context of supranational institutions and ad hoc transnational arrangements.

To be sure, it remains possible for critical opinion to contest claims about the singularity of the moment and seek analogies across time and space. Emergency rule need not escape critical appraisal: the framing on which it relies can be challenged. But to the extent that established agents are able to shape the meaning of events in this way, they wield a significant power. The effects, one may add, will not be confined to the policies directly associated with emergency rule. A focus on exceptional circumstances in one domain easily crowds out consideration of issues elsewhere. When circumstances are cast as unique, it is difficult to keep them in proportion, for who can say what the proper proportions of a unique challenge are? Political discourse becomes consumed with one issue, leaving the scrutiny of others to suffer.

Such observations can be extended with caution to Brexit. Though perhaps shaped less by EU dynamics than domestic ones—the idiosyncrasies of Westminster, its parties and concerns—some similar patterns were involved. From the initial decision to call a referendum on EU membership, some of the enduring questions of British and European politics were remade as the concerns of a particular episode. How the economy might be organized, and how borders should be configured, were perennial questions now recast as distinct. Absorbed into a process of transnational negotiation, with its unfamiliar arrangements and peculiar neologisms, quotidian concerns were reformatted as exceptional, one-off

[23] This was the basis of Schmitt's distinction between laws and 'measures', the latter being the bedrock of restorative emergency rule (*Dictatorship*, p. 8, pp. 210ff.).

[24] Conversely, states in the process of being founded, or 'in transition' from one regime to another (e.g. communism to capitalism), will be prone to similar dynamics as transnational associations.

questions of 'leave' or 'remain', to be settled as a package in a compressed time-frame. No doubt the UK government got more than it bargained for with the 2016 referendum, encountering a public keen to make itself heard, but in the period thereafter its leaders sought to capitalize on the opportunities presented, using ideas of exceptional circumstance to justify secrecy and the push for their preferred settlement. While Labour in opposition sought to resist this reduction of politics to the episodic view, and in particular to challenge its emerging identities of 'remainer' and 'leaver',[25] by others the same perspective was readily embraced, with the emergence of such groups as the Brexit Party for whom all was to be viewed through the prism of the moment.

A theme of this book has been that the politics of emergency is consequential not just in its most intense phases but that it leaves an enduring legacy. We have explored this possibility in terms of its implications for constitutional norms, as well as for popular rejoinders. Reflecting on exceptionalism as a politics of singularities reaffirms this. To govern in ad hoc ways is to reinforce an idea of political institutions as essentially reactive and subordinate to demands that issue elsewhere. It is to approach decision-making as about finding fixes rather than setting normative priorities. Rather than as part of an autonomous framework, political interventions take on the character of instruments for adapting to necessity—as expressions of *doing* rather than *agency*—with longer-term implications for democracy and power.

In addition to a general influence on political culture, exceptionalism has implications for the direction of later policy-making. Consider the Euro crisis once more. To the extent that decisions under the sign of emergency are rooted in immediate concerns of stability rather than principle, revisiting them at a later date will seem risky. Are the ECB's powers likely to be reined in? Being tightly linked to extreme conditions, and held to be necessary to stave off threats, such innovations are difficult to revisit without re-evoking the spectre of crisis. As we have seen, it may also be feared that resisting them puts in doubt the securitizing logic by which they were introduced, thus highlighting the degree of choice that was exercised.

Not all emergency measures are meant to last. Some are intended to mark new beginnings, and are expressed in quasi-constitutional terms to reflect this, while others are ostensibly temporary, to be withdrawn when conditions permit. Some measures, to recall the Schmittian vocabulary, are intended to herald a new order, on the model of 'sovereign dictatorship', while others are to restore the existing order, on the model of 'commissarial dictatorship'.[26] Many more may inhabit an ambiguous space in between. But in all cases the implication is that undoing these

[25] https://www.theguardian.com/politics/2019/may/09/brexit-leave-remain-labour-jeremy-corbyn-campaign-launch-european-elections

[26] Schmitt, *Dictatorship*.

decisions may be hazardous, and if to be done at all then as another exercise of contextual judgement guided by an assessment of functional effects. The anti-democratic features of emergency rule in its moments of greatest intensity are thereby passed on to the future.[27]

Some have argued the EU displays a temporal scheme rather different from that of an independent state. Whereas the latter in its democratic form rests, ideal-typically, on the revisability and reversibility of decisions, the former tends towards an expanding body of presumptively irreversible policies. The difficulty of unmaking treaty commitments once achieved, the reliance on non-majoritarian institutions, and the centrality of deadlines as coordinating devices, are some of the features of transnational politics cited.[28] Reflecting on the politics of emergency, we may add two further observations. One is that the practices of emergency rule magnify this problem, as one of the contexts in which problematic measures have tended to get passed. By generating, in the face of threats, a moment of consensus for initiatives that might otherwise be resisted, the range of policies entrenched is expanded. Often by instrumentalizing the very deadlines that EU politics creates, executives have instituted a range of new binding commitments. Second, the logic of emergency compounds the irreversibility of these, cultivating the notion that to undo them would be to revert to a crisis. It offers rationalization for the introduction of policy measures unaccompanied by a timetable for their revision.

Emergency politics contributes to the fashioning of politics as constitutional politics. Partly this is because anything raised to the level of a major threat is being treated in constitutional terms, as something that goes to the very existence of the framework. Partly too it is because exceptionalism always raises constitutional questions of who has the authority to engage in it, hence its relation to questions of sovereignty. But it is also because it encourages efforts to lock in certain policies, on the understanding that anything less would leave the polity exposed. It gives reason to shift further fields of policy to the transnational level, where they tend to become permanent, even if selectively enforced. In such ways the politics of emergency serves to rationalize features that have long been true of the EU, as well as pushing it further in the same direction.

Emergency Europe

Sometime in the mid-2010s it became common to speak of the *multiple* crises afflicting the EU and its constituent parts. Migration, terrorism, and a variety of

[27] See Oren Gross, 'What "Emergency" Regime?', *Constellations* 13 (1) (2006), pp. 75ff. on the indefinite persistence of emergency measures.

[28] Magnus Ekengren, *The Time of European Governance* (Manchester: Manchester University Press, 2002).

geopolitical questions were added to enduring concerns of an economic kind. Although applying the same word 'crisis' suggested a certain equivalence, what was generally taken to be common was that each involved circumstances qualitatively different from what preceded them and could be expected to follow. The portrayal of political reality as a linear series of unique situations is, I suggest, the logical conclusion of the emergency paradigm we have been examining.

Postulating the exceptionality of the moment is one way to depoliticize it. It obstructs the making of links across different domains, and their linkage to wider historical patterns. In particular it obstructs the application of wider principles. Political reality takes on the appearance of an aggregate of disparate cases and unique demands, each to be handled with regard to functional concerns. Europe's crisis politics of the 2010s addressed itself to socio-economic challenges long in the making, rarely a sudden departure from pre-existing trends and in no meaningful sense resolved. The declining hold of political institutions over accumulations of private debt, or the increasing displacement of persons across borders, are tendencies bound up in endemic and ongoing factors, and ones that interlace with each other. The exceptionalist imaginary, with its emphasis on the sudden and unexpected, obscures this wider context, just as it invites premature moves to declare such episodes concluded.

One might see this as the weakening of the processual view, of the perspective in which developments are grasped in their relations. Such a tendency is not, to be sure, just the effect of a particular governing mode. Emergency politics reinforces this outlook, but has also been enabled by its wider diffusion. As the contemporary anthropologist Jane Guyer observes, societies today seem increasingly attentive to such notions of temporal uniqueness. In a range of contexts she discerns 'rising awareness of a time that is punctuated rather than enduring: of fateful moments and turning points, the date as event rather than as position in a sequence or a cycle, dates as qualitatively different rather than quantitatively cumulative'.[29] The faltering of established ideologies, the embrace of governance methods, and the socio-economic shifts that go with these, are all tendencies that point in this direction. The practices and discourses of emergency redouble them, sharpening their material and symbolic form.

Each application of such an outlook encourages subsequent developments to be approached in the same way. There is a paradox of a kind here. On the one hand, the politics of emergency involves emphasizing what is particular to a certain conjuncture and querying the possibility of comparison. It is to govern a set of circumstances as though they are singular. On the other hand, the legacy of the

[29] Jane Guyer, 'Prophecy and the Near Future: Thoughts on Macroeconomic, Evangelical, and Punctuated Time', *American Ethnologist* 34 (3) (2007), p. 416. Contrast with Arendt's reading of an earlier phase of modernity, when persons 'consider everything in terms of processes and are not concerned with single entities or individual occurrences and their special separate causes': Hannah Arendt, *Between Past and Future* (London: Penguin, 1961), p. 61.

purportedly unique episode is to disrupt expectations of cross-temporal continuity and cultivate anticipation of more such episodes. Each invocation of exceptional circumstances fosters the thought they might reappear in future, in the same domain or another, and prepares the ground thereby for future exercises of discretion oriented to contextual considerations. Despite the denial of substantive continuities, emergency politics amounts to a transferable and recurring template.

In an important work of social theory, Andreas Reckwitz suggests that contemporary societies are increasingly drawn, across many domains, towards an emphasis on particularity over generality. The social world is sketched increasingly by its inhabitants in terms of features that are unique and incomparable—increasingly, he suggests, as a world of singularities.[30] Reckwitz introduces these categories to focus primarily on cultural processes of delineation, e.g. in the self-definition of groups and individuals, or in the production of objects of affect. But the terms capture well the political currents we have been considering, and perhaps even hint at a cultural basis.

The politics of emergency is, on the one hand, a mode of governing where the logic of generalization remains present. It involves taking certain values, interests, and commitments that are said to be imperilled and casting them as the proper concern of all. It entails interpolating a community bound up in the problems at hand, all of them 'in it together', and overlooking the differences of experience and outlook that might otherwise be recognized. But at the same time, it rests very much on the logic of *singularization*, whereby exceptional features of the situation are emphasized over continuities and made the basis for governing measures that reaffirm its distinctiveness. And although they may not be the driving factors in play, the cultural processes Reckwitz describes are arguably connected. In societies attuned to the value of the particular, there will always be a willing constituency for the thought that these are distinctive times, that ours is a special generation, living at a key moment in history. In addition to the many other things that it does, emergency rule perhaps flatters its audience with the suggestion that it lives at a critical juncture.

But what if these *are* distinctive times, one may ask, and the situations faced today are meaningfully unprecedented? Can one so easily discount the thought that there are contexts where a measure of exceptionalism is appropriate? Though the ideas and practices of emergency rule can be abused, are there not legitimate applications—particularly in the areas that EU policy might touch on?

I have argued that one should be wary of seeking standards for exceptionalism, given the licence they provide for a practice easily abused, but a reasonable view

[30] Andreas Reckwitz, *Die Gesellschaft der Singularitäten—Zum Strukturwandel der Moderne* (Berlin: Suhrkamp, 2017), p. 11: 'In der Spätmoderne findet ein gesellschaftlicher Strukturwandel statt, der darin besteht, dass die soziale Logik des Allgemeinen ihre Vorherrschaft verliert an die *soziale Logik des Besonderen*. Dieses Besondere, das Einzigartige, also das, was als nichtaustauschbar und nichtvergleichbar erscheint, will ich mit dem Begriff der Singularität umschreiben.'

still holds that we live in a world of unfamiliar challenges that may plausibly be treated as unprecedented. Beyond the socio-economic problems examined, climate change would be one candidate, clearly transnational in scope. Such challenges, credibly new, threatening and fast-moving, might seem good reason to take the exceptionalist outlook more seriously, a reminder perhaps of the experimentalism needed in governing circles today. Do they not suggest that the problem with emergency thinking in the contexts examined is simply that it devalues a political register we will ultimately be reliant on?

Even when faced with genuinely unfamiliar challenges, the preferable approach seems to connect them to historical patterns, normative principles, and procedural forms. Merit lies in highlighting the continuities as well as the novelties, the analogies that offer points of orientation and suggest possible choices. It lies in seeking connections not just in space but in time, so as to harness existing political structures and normative resources. Advocates of case-centred reasoning remind us that it can be an historically minded exercise, based on looking for analogies and seeking principled consistency across time. Even if the relevant principles are determined through analysis of the problem at hand, the aim is to find continuities rather than highlight the particularities.[31] And it can be an explicitly normative outlook, centred on how to exercise agency in challenging circumstances rather than disclaiming it in the name of necessity.

If situated judgement of this kind offers one model for how extreme circumstances may be governed in principled fashion, the challenge is to embed it in democratic processes. Case-centred reasoning offers little to the legitimation of power if it is exercised by decision-makers privately, in technocratic and informal forums, or in other ways that make it hard for a wider public to contest. Equally, it offers little if exercised publicly and inclusively but by relatively powerless agents unable to connect reasoning processes to what is actually decided.

Political parties have been, and arguably still are, the organizational form in which the challenges of a particular moment can be connected to a longer tradition of analysis and agency. Political ideology has been the basis on which new problems are integrated into an historical perspective and made intelligible in the light of overarching principles. Given the weakness of ideas-based parties in today's world—in the transnational context especially, but also more generally—it seems natural that one should see reversion to a politics of singularities. The strengthening of transnational partisanship, difficult and against the grain as this may be, still looks indispensable for countering this in the EU sphere. It is transnational parties, as associations that aim publicly to connect unfamiliar challenges in uncharted settings to enduring principles, and to do so while acknowledging procedural restraints, that offer the best hope of the judgement

[31] Cf. Sunstein (*Legal Reasoning*), pp. 65ff. on casuistry; on conjunctural analysis, see e.g. Stuart Hall and Doreen Massey, 'Interpreting the Crisis', *Soundings* 44, Spring (2010), pp. 57–71.

and agency needed in changing times. To the extent that coming years produce genuinely unfamiliar challenges, these are the kind of agents one might look to.

Criticizing executive discretion under the sign of emergency should not be mistaken for defending policies and institutions as they stand. On the contrary, as presently configured the EU seems hard to sustain. Rather than to inhibit change, the appropriate aim would be to move away from reliance on the logic of emergency to effect it. It would be to abstain from innovations that depend for their introduction on the suspension of political and legal norms, on casting decisions as measures of last resort, and on concentrating power in small circles of executive elites. When policy-makers rely on the artificial consensus generated by perceptions of threat to get things done, they work on a precarious and dubious basis. To the extent that existing institutions are inadequate, consent for change will need to be pursued by more openly normative and participatory means, as movement-parties of the left contesting both national and European elections have recently sought to do. Their successes have been modest so far, but an indication of what it might mean to put transnational politics on a different footing.

When executive agents are disposed to embrace the emergency mode, there may be little that can ultimately obstruct it. There are no simple fixes, and one should be sceptical that it is a matter of 'getting the policies right'. Rewriting the rules of Eurozone economics and EU migration-response may be desirable moves, but they will not spell an end to the challenges of exceptionalism. Socio-economic conditions evolve, along with societal demands, and there will be legitimate cause for revisions and reversals—something the EU has been badly able to accommodate. More promising would be the construction of a powerful, unitary, and elected EU executive, able to pursue far-reaching changes while constrained by the gaze of an organized public. A version of party democracy on a transnational scale would seem a good place to go, distant as the prospect may be.

There are reasons though why the current structure persists as it does, and they reflect wider changes in executive power. While the politics of emergency finds a receptive setting in the structures of the EU, it is hardly a peculiarity of the transnational sphere. The patterns examined are the highest form of those that have come to pervade national politics in much of the world, as established executive agents align themselves with powerful figures of the socio-economic sphere and aim to govern populations amongst whom their authority is increasingly questioned. Political institutions appear ever less autonomous in relation to the former, ever more so in relation to the latter. Rather than a distinct failing of the EU sphere, the politics of emergency on a cross-border scale is how an array of executive agents, national as well as supranational, have sought to contend with their predicament. It is the remedy sought for problems of national as well as supranational authority, something the rise of the Far Right tracks but does not alter. To the fore comes a form of emergency rule rather different from the

Schmittian image of the strong executive imposing itself by decisive intervention. It is an expression of weakness rather than strength, and the location of sovereignty remains unresolved.

Emergency Europe sees questions of principle recast as a succession of particularities and functional demands, a series of measures of last resort. Its politics stays reactive in nature, bound up in the details of certain moments rather than a basis for setting priorities for the future. For as long as the EU is forged in crisis management, and taken to be the sum of the solutions adopted to this end, it will remain barely distinguishable from the socio-economic forces that many have hoped it would tame.

Bibliography

Ackerman, Bruce (1991), *We the People: Foundations* (Cambridge, MA: Harvard University Press).

Ackerman, Bruce (2006), *Before the Next Attack: Preserving Civil Liberties in an Age of Terrorism* (New Haven: Yale University Press).

Adey, Peter, Ben Anderson, and Stephen Graham (2015), 'Introduction. Governing Emergencies: Beyond Exceptionality', *Theory, Culture & Society* 32 (2) 3–17.

Agamben, Giorgio (2005), *State of Exception* (Chicago: Chicago University Press).

Agamben, Giorgio (2013), 'The Endless Crisis as an Instrument of Power', 4 June https://www.versobooks.com/blogs/1318-the-endless-crisis-as-an-instrument-of-power-in-conversation-with-giorgio-agamben

Akin, William (1977), *Technocracy and the American Dream: The Technocrat Movement, 1900–1941* (Berkeley: University of California Press).

Altman, David (2011), *Direct Democracy Worldwide* (Cambridge: Cambridge University Press).Anderson, Ben (2010), 'Preemption, Precaution, Preparedness: Anticipatory Action and Future Geographies', *Progress in Human Geography* 34(6): 777–98.

Anderson, Ben (2017), 'Emergency Futures: Exception, Urgency, Interval, Hope', *Sociological Review* 65 (3), 463–77.

Anderson, Christopher J., André Blais, Shaun Bowler, Todd Donovan, and Ola Listhaug (2005), *Losers' Consent: Elections and Democratic Legitimacy* (Oxford: Oxford University Press).

Anzia, Sarah F. (2013), *Timing and Turnout: How Off-cycle Elections Favor Organized Groups* (Chicago: University of Chicago Press).

Aradau, Claudia (2004), 'Security and the Democratic Scene: Desecuritization and Emancipation', *Journal of International Relations and Development* 7, 388–413.

Arendt, Hannah (1961), *Between Past and Future* (London: Penguin).

Arendt, Hannah (1990), *On Revolution* (London: Penguin).

Asmussen, Jörg (2012), 'Stability Guardians and Crisis Managers: Central Banking in Times of Crisis and Beyond', Lecture at the Goethe-Universität Frankfurt, 11 September (https://www.ecb.europa.eu/press/key/date/2012/html/sp120911.en.html).

Auer, Štefan (2019), 'Merkel's Germany and the European Union: Between Emergency and the Rule of Rules', *Government and Opposition* 1–19.

Balzacq, Thierry, Sarah Léonard, and Jan Ruzicka (2016), '"Securitization" Revisited: Theory and Cases', *International Relations* 30 (4), 494–531.

Baquerro-Cruz, Julio (2012), 'Legal Pluralism and Institutional Disobedience in the European Union', in Matej Avbelj and Jan Komárek (eds), *Constitutional Pluralism in the European Union and Beyond* (Oxford: Hart).

Barroso, José Manuel (2011), Statement to the press following the meeting of eurozone Heads of State or Government, Brussels, 27 October, speech/11/713.

Barroso, José Manuel (2014), Valedictory speech by Commission President to the European Parliament, 21 October, Strasbourg.

Bauer, Michael and Stefan Becker (2014), 'The Unexpected Winner of the Crisis: The European Commission's Strengthened Role in Economic Governance', *Journal of European Integration* 36 (3), 213–29.

Bauman, Zygmunt (2004), 'From Social State to Security State', in Bauman, *Europe: An Unfinished Adventure* (Cambridge: Polity).

Beck, Ulrich (2009), 'Critical Theory of World Risk Society' *Constellations*, 16 (1), 3–22.

Beck, Ulrich (2013), *German Europe* (Cambridge: Polity).

Becker, Stefan, Michael W. Bauer, Sara Connolly, and Hussein Kassim (2016), 'The Commission: Boxed In and Constrained, but Still an Engine of Integration', *West European Politics* 39 (5), 1011–31.

Beckert, Jens (2016), *Imagined Futures: Fictional Expectations and Capitalist Dynamics* (Cambridge MA: Harvard University Press).

Bellamy, Richard (2001), 'The Rule of Law and the Rule of Persons', *Critical Review of International Social and Political Philosophy* 4 (4), 221–51.

Benz, Arthur (2003), 'Compounded Representation in EU Multi-level Governance', in Beate Kohler-Koch (ed.), *Linking EU and National Governance* (Oxford: Oxford University Press), 82–110.

Best, Jacqueline (2018), 'Technocratic Exceptionalism: Monetary Policy and the Fear of Democracy', *International Political Sociology* 12 (4), 328–45.

Beukers, Thomas, Bruno de Witte, and Claire Kilpatrick (eds) (2017), *Constitutional Change through Euro-Crisis Law* (Cambridge: Cambridge University Press).

Bevan, Gwyn and Christopher Hood (2006), 'What's Measured Is What Matters: Targets and Gaming in the English Public Health Care System', *Public Administration* 84 (3), 517–38.

Bhambra, Gurminder K. (2017), 'The Current Crisis of Europe: Refugees, Colonialism, and the Limits of Cosmopolitanism', *European Law Journal* 23 (5), 395–405.

Bickerton, Christopher J. (2012), *European Integration: From Nation-States to Member-States* (Oxford: Oxford University Press).

Bickerton, Christopher J. and Carlo Invernizzi Accetti (2017), 'Populism and Technocracy: Opposites or Complements?', *Critical Review of Social and Political Philosophy* 20 (2), 186–206.

Bickerton, Christopher J., Dermot Hodson, and Uwe Puetter (2015), 'The New Intergovernmentalism: European Integration in the Post-Maastricht Era', *Journal of Common Market Studies* 53 (4), 703–22.

Biebricher, Thomas (2013), 'Europe and the Political Philosophy of Neoliberalism', *Contemporary Political Theory* 12, 338–75.

Biebricher, Thomas (2017), 'Ordoliberalism as a Variety of Neoliberalism', in Josef Hien and Christian Joerges (eds), *Ordo-Liberalism, Law and the Rule of Economics* (Oxford: Hart), 103–113.

Bigo, Didier (2002), 'Security and Immigration: Toward a Critique of the Governmentality of Unease', *Alternative* 27 (Special Issue), 63–92.

Blyth, Mark (2013), *Austerity: The History of a Dangerous Idea* (Oxford: Oxford University Press).

Blyth, Mark and Richard Katz (2005), 'From Catch-all Politics to Cartelisation: The Political Economy of the Cartel Party', *West European Politics* 28 (1), 33–60.

Böckenförde, Ernst-Wolfgang (2010), 'Kennt die europäische Not kein Gebot?', *Neue Zürcher Zeitung*, 21 June (https://www.nzz.ch/kennt_die_europaeische_not_kein_gebot-1.6182412).

Böhm, Franz (1966), 'Privatrechtsgesellschaft und Marktwirtschaft', *ORDO Jahrbuch für die Ordnung von Wirtschaft und Gesellschaft* 17, 75–152.

Boin, Arjen, Magnus Ekengren, and Mark Rhinard (2013), *The European Union as Crisis Manager: Patterns and Prospects* (Cambridge: Cambridge University Press).

Boin, Arjen, Paul 't Hart and Allan McConnell (2009), 'Crisis Exploitation: Political and Policy Impacts of Framing Contests', *Journal of European Public Policy* 16 (1).
Boltanski, Luk (2011), *On Critique* (Cambridge: Polity).
Bonefeld, Werner (2017), *The Strong State and the Free Economy* (Lanham: Rowman & Littlefield).
Braunthal, J. (1960), *History of the International, vol. I, 1864–1914* (New York: Praeger).
Brennan, Geoffrey, Lina Eriksson, Robert E. Goodin, and Nicholas Southwood (2013), *Explaining Norms* (Oxford: Oxford University Press).
Brown, Wendy (2015), *Undoing the Demos: Neoliberalism's Stealth Revolution* (MIT: Zone Books).
Brubaker, Rogers (2017), 'Why Populism?', *Theory and Society* 46 (5), 357–85.
Bruff, Ian (2014), 'The Rise of Authoritarian Neoliberalism', *Rethinking Marxism: A Journal of Economics, Culture and Society* 26 (1), 113–29.
Brunkhorst, Hauke (2014), *Das doppelte Gesicht Europas. Zwischen Kapitalismus und Demokratie* (Frankfurt/Main: Suhrkamp).
Brunkhorst, Hauke (2016), 'The European Dual State: The Double Structural Transformation of the Public Sphere and the Need for Repoliticization', in Jiří Přibáň (ed.), *Self-Constitution of European Society: Beyond EU Politics, Law and Governance* (London: Routledge), 239–73.
Bulmer, Simon, and Wolfgang Wessels (1987), *The European Council. Decision-making in European Politics* (London: Macmillan).
Buzan, Barry and Ole Wæver (2009), 'Macrosecuritization and Security Constellations: Reconsidering Scale in Securitization Theory', *Review of International Studies*, 35 (2), 253–76.
Buzan, Barry, Ole Wæver and Jaap de Wilde (1998), *Security: A New Framework for Analysis*, (Boulder CO: Lynne Rienner).
Calhoun, Craig (2004), 'A World of Emergencies: Fear, Intervention and the Limits of Cosmopolitan Order', *Canadian Review of Sociology* 41 (4), 373–95.
Campesi, Giuseppe (2018), 'Crisis, Migration and the Consolidation of the EU Border Control Regime', *International Journal of Migration and Border Studies* 4 (3), 196–221.
Canovan, Margaret (2005), *The People* (Cambridge: Polity).
Caramani, Daniele (2017), 'Will vs Reason: The Populist and Technocratic Forms of Political Representation and Their Critique to Party Government', *American Political Science Review* 111 (1), 54–67.
Castelli Gattinara, Pietro (2017), 'The "Refugee Crisis" in Italy as a Crisis of Legitimacy', *Contemporary Italian Politics* 9 (3), 318–31.
Celikates, Robin (2016), 'Democratizing Civil Disobedience', *Philosophy and Social Criticism* 42 (10), 982–94.
Celikates, Robin (2016), 'Rethinking Civil Disobedience as a Practice of Contestation—Beyond the Liberal Paradigm', *Constellations* 23 (1), 37–45.
Centeno, Miguel Angel (1993), 'The New Leviathan: The Dynamics and Limits of Technocracy', *Theory and Society* 22 (3), 307–35.
Chalmers, Damian (2016), 'Crisis Reconfiguration of the European Constitutional State', in Chalmers, Markus Jachtenfuchs, and Christian Joerges (eds), *The End of the Eurocrats' Dream: Adjusting to European Diversity* (Cambridge: Cambridge University Press), 266–98.
Chalmers, Damian and Barroso, Luis (2014), 'What *Van Gend en Loos* Stands For', *International Journal of Constitutional Law*, 12 (1), 105–34.
Chalmers, Damian, Gareth Davies, and Giorgio Monti (2014), *European Union Law: Text and Materials* (Cambridge: Cambridge University Press).

Cheneval, Francis and Mónica Ferrín (2018), 'Referendums in the European Union: Defective by Birth?', *Journal of Common Market Studies* 56 (5), 1178–94.

Clift, Ben (2018), *The IMF and the Politics of Austerity in the Wake of the Global Financial Crisis* (Oxford: Oxford University Press).

Codogno, Lorenzo, Guillaume Odinet, and Flavio Padrini (2009), 'The Use of Targets in the Lisbon Strategy', *Rivista di Politica Economica* January–March, 3–21.

Cohen, Elizabeth (2018), *The Political Value of Time* (Cambridge: Cambridge University Press).

Cohen, Morris R. (1914), 'Rule vs Discretion', *Journal of Philosophy, Psychology and Scientific Methods* 11, 208–15.

Collins, Harry and Robert Evans (2007), *Rethinking Expertise* (Chicago: University of Chicago Press).

Côté, Adam (2016), 'Agents without Agency: Assessing the Role of the Audience in Securitisation Theory', *Security Dialogue* 47 (6), 541–58.

Curley, Tyler M. (2015), 'Models of Emergency Statebuilding in the United States', *Perspectives on Politics* 13 (3), 697–713.

Curtin, Deidre (2009), *Executive Power of the European Union. Law, Practices and the Living Constitution* (Oxford: Oxford University Press).

Curtin, Deidre (2014), 'Challenging Executive Dominance in European Democracy', *Modern Law Review* 77 (1), 1–32.

Davies, William (2014), *The Limits of Neoliberalism: Authority, Sovereignty and the Logic of Competition* (London: Sage).

Davis Cross, Mai'a (2017), *The Politics of Crisis in Europe* (Cambridge: Cambridge University Press).

Davis Cross, Mai'a and Xinru Ma (2015), 'EU crises and integrational panic: the role of the media', *Journal of European Public Policy* 22 (8), 1053–1070.

Davitti, Daria (2019), 'Biopolitical Borders and the State of Exception in the European Migration "Crisis"', *European Journal of International Law* 29 (4), 1173–96.

Dawson, Mark (2015), 'The Legal and Political Accountability Structure of "Post-Crisis" EU Economic Governance', *Journal of Common Market Studies* 53 (5), 976–93.

D'Eramo, Marco (2013), 'Populism and the New Oligarchy', *New Left Review* 82 (July/August), 5–28.

Devlin, John F. (1991), 'The Baath Party: Rise and Metamorphosis', *The American Historical Review* 96 (5), 1396–1407.

De Wilde, Marc (2012), 'The Dictator's Trust: Regulating and Constraining Emergency Powers in the Roman Republic', *History of Political Thought* 33 (4), 555–77.

De Wilde, Pieter and Michael Zürn (2012), 'Can the Politicization of European Integration be Reversed?', *Journal of Common Market Studies* 50 (S1), 137–53.

Dewey, John (1958), 'The Natural History of Form', in Dewey, *Art as Experience* (New York: Capricorn), 134–62.

Diab, R. (2015), *The Harbinger Theory: How the Post-9/11 Emergency Became Permanent and the Case for Reform* (Oxford: Oxford University Press).

DiEM25's July 2017 'Guide to Constructive Disobedience': https://www.opendemocracy.net/can-europe-make-it/diem25/guide-to-constructive-disobedience

Dinan, Desmond (2014), 'Governance and Institutions: The Unrelenting Rise of the European Parliament', *Journal of Common Market Studies* 52, Annual Review, 109–24.

Dinan, Desmond (2016), 'Governance and Institutions: A More Political Commission' *Journal of Common Market Studies* 54, Annual Review, 101–16.

Duchêne, Francois (1994), *Jean Monnet: The First Statesman of Interdependence* (New York: W. W. Norton & Company).

Dyson, Kenneth (2013), 'Sworn to Grim Necessity? Imperfections of European Economic Governance, Normative Political Theory, and Supreme Emergency', *Journal of European Integration* 35 (3), 207–22.

Dyzenhaus, David (2006), *The Constitution of Law: Legality in a Time of Emergency* (Cambridge: Cambridge University Press).

Eger, Thomas, Stefan Oeter, and Stefan Voigt (2017), *International Law and the Rule of Law under Extreme Conditions: An Economic Perspective. Contributions to the XIVth Travemünde Symposium on the Economic Analysis of Law (March 27–29, 2014)* (Tübingen: Mohr Siebeck).

Eichengreen, Barry (2015), *Hall of Mirrors: The Great Depression, The Great Recession, and the Uses-and Misuses-of History* (Oxford: Oxford University Press).

Ekengren, Magnus (2002), *The Time of European Governance* (Manchester: Manchester University Press).

Ellman, Michael (2010), 'Regional Influences on the Formulation and Implementation of NKVD Order 00447', *Europe-Asia Studies* 62 (6), 915–31.

Elster, Jon (2004), 'Comments on the Paper by Ferejohn and Pasquino', *I•CON* 2 (2), 240–3.

Enroth, Henrik (2014), 'Governance: The Art of Governing after Governmentality', *European Journal of Social Theory* 17 (1), 60–76.

European Central Bank (2014), 'Exchange of Views of Benoît Cœuré with ECON on Troika Matters: Introductory Remarks by Benoît Cœuré, Member of the Executive Board of the ECB', Brussels, 13 February (https://www.ecb.europa.eu/press/key/date/2014/html/sp140213.en.html).

European Central Bank (2014), Replies to the Questionnaire of the European Parliament Supporting the Own Initiative Report Evaluating the Structure, the Role and Operations of the 'Troika' (Commission, ECB and the IMF) Actions in Euro Area Programme Countries, 2014 (https://www.ecb.europa.eu/pub/pdf/other/140110_ecb_response_troika_questionnaireen.pdf).

European Commission (2012), 'A Blueprint for a Deep and Genuine Economic and Monetary Union: Launching a European Debate', COM(2012) 777 final/2, 28th November (Brussels: European Commission).

European Commission (2013), Response to the European Parliamentary 'Questionnaire Supporting the Own Initiative Report Evaluating the Structure, the Role and Operations of the "Troika" (Commission, ECB and the IMF) Actions in Euro Area Programme Countries'.

European Commission (2015), 'Communication from the Commission to the European Parliament and the Council: A European Border and Coast Guard and Effective Management of Europe's External Borders' (Brussels: EC).

European Commission (2015), 'Completing Europe's Economic and Monetary Union' [Five Presidents' Report'], Report by Jean-Claude Juncker, in close cooperation with Donald Tusk, Jeroen Dijsselbloem, Mario Draghi, and Martin Schulz (Brussels: EC).

European Commission (2015), 'A European Border and Coast Guard to Protect Europe's External Borders', press release, 15 December (Brussels: EC).

European Commission (2016), 'Proposal for a Council Implementing Decision Setting Out a Recommendation for Temporary Internal Border Control in Exceptional Circumstances Putting the Overall Functioning of the Schengen Area at Risk' (Brussels: EC).

European Community (1974), 'Final Communiqué of the Meeting of Heads of Government of the Community, Paris, 9–10 December 1974', in *Bulletin of the European Communities* 12.

European Council (2010), Press release of the Extraordinary Council meeting, Economic and Financial Affairs, Brussels, 9/10 May.

European Court of Auditors (2015), 'Financial Assistance Provided to Countries in Difficulties', Special Report 18: http://www.eca.europa.eu/Lists/ECADocuments/SR15_18/SR_CRISIS_SUPPORT_EN.pdf).

European Parliamentary Research Service (2016), 'The European Council and Crisis Management', PE 573.283, February.

European University Institute (EUI) (2014), 'Constitutional Change through Euro Crisis Law: Austria', http://eurocrisislaw.eui.eu.

European University Institute (EUI) (2014), 'Constitutional Change through Euro Crisis Law: Cyprus', http://eurocrisislaw.eui.eu.

European University Institute (EUI) (2014), 'Constitutional Change through Euro Crisis Law: France', http://eurocrisislaw.eui.eu.

European University Institute (EUI) (2014), 'Constitutional Change through Euro Crisis Law: Greece', http://eurocrisislaw.eui.eu.

European University Institute (EUI) (2014), 'Constitutional Change through Euro Crisis Law: Italy', http://eurocrisislaw.eui.eu.

European University Institute (EUI) (2014), 'Constitutional Change through Euro Crisis Law: Netherlands', http://eurocrisislaw.eui.eu.

European University Institute (EUI) (2014), 'Constitutional Change through Euro Crisis Law: Portugal', http://eurocrisislaw.eui.eu.

European University Institute (EUI) (2014), 'Constitutional Change through Euro Crisis Law: Romania', http://eurocrisislaw.eui.eu.

European University Institute (EUI) (2014), 'Constitutional Change through Euro Crisis Law: Slovenia', http://eurocrisislaw.eui.eu.

European University Institute (EUI) (2014), 'Constitutional Change through Euro Crisis Law: Spain', http://eurocrisislaw.eui.eu.

Ferejohn, John and Pasquale Pasquino (2004), 'The Law of Exception: A Typology of Emergency Powers,' *International Journal of Constitutional Law* 2.

Ferrara, Alessandro (2008), *The Force of the Example* (New York: Columbia University Press).

Fischer-Lescano, Andreas (2014), *Human Rights in Times of Austerity Policy: The EU Institutions and the Conclusion of Memoranda of Understanding* (Baden-Baden: Nomos).

Fontaine, Pascal (1979), 'Jean Monnet's Role in the Birth of the European Council' (September), translation of Fontaine (1979), 'Le rôle de Jean Monnet dans la genèse du Conseil européen', *Revue du Marché commun* (Sept.) 229, available at: https://www.cvce.eu/content/publication/2006/12/7/ad29595e-0b0a-49b7-ae65-c41c54344a41/publishable_en.pdf

Fossum, John Erik and Agustín José Menéndez (2014), *The European Union in Crises or the European Union as Crises?*, ARENA Report No 2/14 (Oslo: ARENA).

Fraenkel, Ernst (1941/1917), *The Dual State: A Contribution to the Theory of Dictatorship* (Oxford: Oxford University Press).

Franceschet, Antonio (2015), 'Theorizing State Civil Disobedience in International Politics', *Journal of International Political Theory* 11 (2), 239–56.

Freeden, Michael (2017), 'After the Brexit Referendum: Revisiting Populism as an Ideology', *Journal of Political Ideologies* 22 (1), 1–11.

Friedman, Milton (1948), 'A Monetary and Fiscal Framework for Economic Stabilisation', *American Economic Review* 38, 245–64.

Friedrich, Carl (1968), *Constitutional Government and Democracy: Theory and Practice in Europe and America* (Waltham, Mass: Blaisdell).

Genschel, Philipp and Markus Jachtenfuchs (2018), 'From Market Integration to Core State Powers: The Eurozone Crisis, the Refugee Crisis and Integration Theory', *Journal of Common Market Studies* 56 (1), 178–96.

Georgakakis, Didier (2010), 'Do Skills Kill? Les enjeux de la requalification de la compétence des eurofonctionnaires', *Revue française d'administration publique* 133, 61–80.

Georgakakis, Didier (2014), 'À la fois pilote et victime de l'austérité ? Les transformations de l'administration de l'Union européenne sous l'effet de la crise economique et financière', *Revue Française d'Administration publique* 151–2, 805–24.

Georgakakis, Didier and Jay Rowell (2013), *The Field of Eurocracy: Mapping EU Actors and Professionals* (Basingstoke: Palgrave).

Gerber, David J. (1994), 'Constitutionalizing the Economy: German Neo-liberalism, Competition Law and the "New" Europe', *The American Journal of Comparative Law* 42 (1), 25–84.

Geuna, Marco (2015), 'Machiavelli and the Problem of Dictatorship', *Ratio Juris* 28 (2), 226–41.

Giddens, Anthony (1990), *The Consequences of Modernity* (Stanford: Stanford University Press).

Goetz, Klaus (2009), 'How Does the EU Tick? Five Propositions on Political Time', *Journal of European Public Policy* 16 (2), 202–20.

Goetz, Klaus (2015), 'Synchronisation demokratischen Regierens in der Europäischen Union', *Leviathan: Berliner Zeitschrift für Sozialwissenschaft* 43 (Sonderband 30), 132–51.

Goodin, Robert E. (1998), 'Keeping Political Time: The Rhythms of Democracy', *International Political Science Review* 19 (1), 39–54.

Grimm, Dieter (2017), *The Constitution of European Democracy* (Oxford: Oxford University Press).

Gross, Oren (2006), 'What "Emergency" Regime?', *Constellations* 13 (1), 74–88.

Gross, Oren (2008), 'Extra-Legality and the Ethic of Political Responsibility', in Victor V. Ramraj (ed.), *Emergencies and the Limits of Legality* (Cambridge: Cambridge University Press), 60–94.

Gross, Oren and Fionnuala Ní Aoláin (2006), *Law in Times of Crisis: Emergency Powers in Theory and Practice* (Cambridge: Cambridge University Press).

Guyer, Jane (2007), 'Prophecy and the Near Future: Thoughts on Macroeconomic, Evangelical, and Punctuated Time', *American Ethnologist* 34 (3), 409–42.

Haas, Ernst B. (1968), *The Uniting of Europe: Politics, Social and Economic Forces, 1950–1957* (Stanford: Stanford University Press).

Habermas, Jürgen (2015), *The Lure of Technocracy*, trans. Ciarin Cronin (Cambridge: Polity).

Hall, Stuart and Doreen Massey (2010), 'Interpreting the Crisis', *Soundings* 44, Spring, 57–71.

Hamilton, Alexander, James Madison, and John Jay (1787/2008), *The Federalist Papers* (Oxford: Oxford University Press).

Hansen, Lene (2012), 'Reconstructing Desecuritisation: The Normative-Political in the Copenhagen School and Directions for How to Apply It', *Review of International Studies* 38 (3), 525–46.

Harmes, Adam (2006), 'Neoliberalism and Multilevel Governance', *Review of International Political Economy* 13 (5), 725–49.

Hay, Colin (2004), 'The Normalizing Role of Rationalist Assumptions in the Institutional Embedding of Neoliberalism', *Economy and Society* 33 (4), 500–27.

Hay, Colin and Ben Rosamond (2002), 'Globalisation, European Integration and the Discursive Construction of Economic Imperatives', *Journal of European Public Policy* 9 (2), 147–67.

Hayek, Friedrich August von (1945), 'The Use of Knowledge in Society', *American Economic Review* 35, 519–30.

Hayek, Friedrich August von (1952), 'The Economic Conditions of Interstate Federalism', in Hayek, *Individualism and Economic Order* (London: Routledge & Kegan Paul).

Hayek, Friedrich August von (1955), *The Counter-Revolution of Science* (New York: Free).

Hayek, Friedrich August von (1973), 'Rules and Order', vol. 1 of *Law, Legislation and Liberty* (London: Routledge & Kegan Paul).

Head, Michael (2015), *Emergency Powers in Theory and Practice: The Long Shadow of Carl Schmitt* (London: Routledge).

Henning, C. Randall (2017), *Tangled Governance: International Regime Complexity, the Troika, and the Euro Crisis* (Oxford: Oxford University Press).

Herman, Lise (2017), 'Democratic Partisanship: From Theoretical Ideal to Empirical Standard', *American Political Science Review* 111 (4), 738–54.

Hien, Josef and Christian Joerges (2017) (eds), *Ordo-Liberalism, Law and the Rule of Economics* (Oxford: Hart).

Hirschman, Albert (1996), *The Passions and the Interests: Political Arguments for Capitalism Before Its Triumph* (Princeton: Princeton University Press).

Hobolt, Sara B. and James Tilley (2016), 'Fleeing the Centre: The Rise of Challenger Parties in the Aftermath of the Euro Crisis', *West European Politics*, 39 (5), 971–91.

Holmes, Stephen (2009), 'In Case of Emergency: Misunderstanding Tradeoffs in the War on Terror', *California Law Review* 97 (2), 301–55.

Honig, Bonnie (2009), *Emergency Politics: Paradox, Law, Democracy* (Princeton: Princeton University Press).

Hooghe, Lisbet and Gary Marks (2009), 'A Postfunctionalist Theory of European Integration: From Permissive Consensus to Constraining Dissensus', *British Journal of Political Science* 39 (1), 1–23.

Hooghe, Lisbet and Gary Marks (2018), 'Cleavage Theory Meets Europe's Crises: Lipset, Rokkan, and the Transnational Cleavage', *Journal of European Public Policy* 25 (1), 109–35.

Imbrogno, Anthony F. (2016), 'The Founding of the European Council: Economic Reform and the Mechanism of Continuous Negotiation', *Journal of European Integration* 38 (6), 719–36.

Independent Evaluation Office of the IMF (2016), 'The IMF and the Crises in Greece, Ireland and Portugal', 8 July (http://www.ieo-imf.org/ieo/pages/CompletedEvaluation267.aspx).

Innerarity, Daniel (2014), 'Does Europe Need a Demos to be Truly Democratic?', *LSE Europe in Question Series (LEQS)*, No. 77.

Invernizzi Accetti, Carlo (2019), *What Is Christian Democracy?* (Cambridge: Cambridge University Press).

Invernizzi Accetti, Carlo and Fabio Wolkenstein (2017), 'The Crisis of Party Democracy, Cognitive Mobilization, and the Case for Making Parties More Deliberative', *American Political Science Review* 111 (1), 97–109.

Ipsen, Hans Peter (1970), *Verfassungsperspektiven der Europäischen Gemeinschaften* (Berlin: de Gruyter).

Isiksel, Turkuler (2016), *Europe's Functional Constitution: A Theory of Constitutionalism Beyond the State* (Oxford, Oxford University Press).

Jacobs, Andreas (2016), 'In Troubled Waters: NATO's New Maritime Activity in the Aegean', NDC Research Report, Research Division, NATO Defense College 1/16 (March).

Jessop, Bob (2009), 'The Spatio-Temporal Dynamics of Globalising Capital and their Impact on State Power and Democracy', in Hartmut Rosa and William Scheuerman (eds.) *High-Speed Society* (University Park: Pennsylvania State University Press), 135–58.

Joerges, Christian (2014), 'Brother, Can You Paradigm?', *I•CON* 12 (3), 769–85.

Joerges, Christian (2014), 'Europe's Economic Constitution in Crisis and the Emergence of a New Constitutional Constellation', in John Erik Fossum and Agustín José Menéndez (2014), *The European Union in Crises or the European Union as Crises?*, ARENA Report No 2/14 (Oslo: ARENA).

Joerges, Christian (2014), 'Law and Politics in Europe's Crisis: On the History of the Impact of an Unfortunate Configuration', *Constellations* 21 (2), 249–61.

Joerges, Christian (2016), 'What Is Left of the European Economic Constitution II? From Pyrrhic Victory to Cannae Defeat', in Poul F. Kjaer and Niklas Olsen (eds), *Critical Theories of Crisis in Europe: From Weimar to the Euro* (Lanham MD: Rowman & Littlefield), 143–60.

Joerges, Christian and Christian Kreuder-Sonnen (2017), 'European Studies and the European Crisis: Legal and Political Science between Critique and Complacency,' *European Law Journal* 23 (1–2), 118–39.

Johansson, Karl Magnus (2017), 'The Role of Europarties in EU Treaty Reform: Theory and Practice', *Acta Politica* 52 (3), 286–305.

Jones, Erik, R. Daniel Kelemen, and Sophie Meunier (2016), 'Failing Forward? The Euro Crisis and the Incomplete Nature of European Integration', 49 (7), 1010–34.

Jones, Reece, Corey Johnson, Wendy Brown, Gabriel Popescu, Polly Pallister-Wilkins, Alison Mountz, and Emily Gilbert (2017), 'Interventions on the State of Sovereignty at the Border', *Political Geography* 59, 1–10.

Jonsen, Albert R. and Stephen Toulmin (1988), *The Abuse of Casuistry: A History of Moral Reasoning* (Berkeley: University of California Press).

Juncker, Jean-Claude (2014), 'A New Start for Europe: My Agenda for Jobs, Growth, Fairness and Democratic Change', Opening Statement in the European Parliament Plenary Session, 15 July (http://www.eesc.europa.eu/resources/docs/jean-claude-juncker—political-guidelines.pdf).

Juncker, Jean-Claude (2015), 'State of the Union 2015: Time for Honesty, Unity and Solidarity', Strasbourg, 9 September 2015 (http://europa.eu/rapid/press-release_SPEECH-15-5614_en.htm).

Kaelble, Hartmut (2014), 'Eine beispiellose Krise? Die Krise der europäischen Integration seit 2010 im historischen Vergleich', in M. Heidenreich (ed.) *Krise der europäischen Vergesellschaftung?* (Wiesbaden: Springer).

Kaiser, Wolfram (2007), *Christian Democracy and the Origins of European Union* (Cambridge: Cambridge University Press).

Kalyvas, Andreas (2007), 'The Tyranny of Dictatorship: When the Greek Tyrant Met the Roman Dictator', *Political Theory* 35 (4), 412–442.

Kalyvas, Andreas (2008), *Democracy and the Politics of the Extraordinary: Max Weber, Carl Schmitt, and Hannah Arendt* (Cambridge: Cambridge University Press).

Katz, Richard, and Peter Mair (2009), 'The Cartel Party Thesis', *Perspectives on Politics* 7 (4), 753–66.

Kauppi, Niilo (2014), 'Knowledge Warfare: Social Scientists as Operators of Global Governance', *International Political Sociology* 8 (3), 330–2.

Kauppi, Niilo and Mikael R. Madsen (2013) (eds), *Transnational Power Elites: The New Professionals of Governance, Law and Security* (London: Routledge).

Kauppi, Niilo and Mikael R. Madsen (2014), 'Fields of Global Governance: How Transnational Power Elites Can Make Global Governance Intelligible', *International Political Sociology* 8 (3), 324–30.

Kelly, Duncan (2017), 'Carl Schmitt's Political Theory of Dictatorship', in Jens Meierhenrich and Oliver Simons (eds), *Oxford Handbook of Carl Schmitt* (Oxford: Oxford University Press), 217–44.

Kelly, Duncan (2017), 'Populism and the History of Popular Sovereignty', in Cristobal Rovira Kaltwasser, Paul Taggart, Paulina Ochoa Espejo, and Pierre Ostiguy (eds), *The Oxford Handbook of Populism* (Oxford: Oxford University Press), 511–34.

Kennedy, David (2016), *A World of Struggle: How Power, Law and Expertise Shape Global Political Economy* (Princeton: Princeton University Press).

Kilpatrick, Claire (2015), 'On the Rule of Law and Economic Emergency: The Degradation of Basic Legal Values in Europe's Bailout', *Oxford Journal of Legal Studies* 35, 325–53.

Kincaid, G. Russell (2016), 'The IMF's Role in the Euro Area Crisis: What Are the Lessons from the IMF's Participation in the Troika?', *Independent Evaluation Office of the IMF*, Background paper BP/16-02/06.

Kinski, Lucy (2018), 'Whom to Represent? National Parliamentary Representation during the Eurozone Crisis', *Journal of European Public Policy* 25 (3), 346–68.

Kjaer, Poul (2017), 'European Crises of Legally-Constituted Public Power: From the "Law of Corporatism" to the "Law of Governance"', *European Law Journal* 23 (5), 417–30.

Klein, Naomi (2007), *The Shock Doctrine: The Rise of Disaster Capitalism* (London: Penguin).

Klein, Naomi (2017), *No Is Not Enough: Defeating the New Shock Politics* (London: Allen Lane).

Klieman, Aaron (1976), 'Emergency Politics: The Growth of Crisis Government', *Conflict Studies* 70, April, 1–19.

Koselleck, Reinhart (2004), *Futures Past: On the Semantics of Historical Time* (New York: Columbia University Press).

Kozicharow, Nicola (2011), 'Filipp Maliavin's Troika Dream of 1933', paper presented at Cambridge University, 15 February.

Krebs, Ronald (2009), 'In the Shadow of War: The Effects of Conflict on Liberal Democracy', *International Organization* 63 (1), 177–210.

Kreuder-Sonnen, Christian (2016), 'Beyond Integration Theory: The (Anti-)Constitutional Dimension of European Crisis Governance', *Journal of Common Market Studies* 54 (6), 1350–66.

Kreuder-Sonnen, Christian (2018), 'An Authoritarian Turn in Europe and European Studies?', *Journal of European Public Policy* 25 (3), 452–64.

Kreuder-Sonnen, Christian (2018), 'Political Secrecy in Europe: Crisis Management and Crisis Exploitation', *West European Politics* 14 (4), 958–80.

Kühnhardt, Ludger (ed.) (2009), *Crises in European Integration: Challenges and Responses, 1945-2005* (Oxford: Berghahn).

Kumm, Mathias (2012), 'The Moral Point of Constitutional Pluralism: Defining the Domain of Legitimate Institutional Civil Disobedience and Conscientious Objection', in Julie Dickson and Pavlos Eleftheriadis (eds), *Philosophical Foundations of EU Law* (Oxford: Oxford University Press), 216–46.

Kydland, Finn E. and Edward C. Prescott (1977), 'Rules rather than Discretion: The Inconsistency of Optimal Plans', *Journal of Political Economy* 85 (3), 473–92.

Lacey, Joseph (2017), *Centripetal Democracy: Democratic Legitimacy and Political Identity in Belgium, Switzerland and the European Union* (Oxford: Oxford University Press).

Laffan, Brigid and Pierre Schlosser (2016), 'Public Finances in Europe: Fortifying EU Economic Governance in the Shadow of the Crisis', *Journal of European Integration* 38 (3), 237–49.

Lakoff, Andrew (2007), 'Preparing for the Next Emergency', *Public Culture* 19 (2), 247–71.

Langenohl, Andreas (2017), 'Securities Markets and Political Securitization: The Case of the Sovereign Debt Crisis in the Eurozone', *Security Dialogue* 48 (2), 131–48.

Lapavitsas, Costas (2012), *Crisis in the Eurozone* (London: Verso).

Lara, María Pía (2008), 'Reflective Judgment as World Disclosure', *Philosophy and Social Criticism* 34 (1–2), 83–100.

Laux, Henning and Hartmut Rosa (2015), 'Clockwork Politics. Fünf Dimensionen politischer Zeit', *Leviathan: Berliner Zeitschrift für Sozialwissenschaft* 43 (Sonderband 30), 52–70.

Lazar, Nomi Claire (2009), *States of Emergency in Liberal Democracies* (Cambridge: Cambridge University Press).

Lemke, Matthias (2018), 'What Does State of Exception Mean? A Definitional and Analytical Approach', *Zeitschrift für Politikwissenschaft* 3 (online early).

Lemoine, Benjamin (2017), 'Measuring and Restructuring the State: Debt Metrics and the Control of Present and Future Political Order', in Desmond King and Patrick Le Galès (2017), *Reconfiguring European States in Crisis* (Oxford: Oxford University Press), 313–30.

Levine, Caroline (2014), *Forms: Whole, Rhythm, Hierarchy, Network* (Princeton: Princeton University Press).

Lewis, Jeffrey (2015), 'The Council of the European Union and the European Council', in José M. Magone (ed.), *Routledge Handbook of European Politics* (Abingdon: Routledge), 219–34.

Lincott, Andrew (2003), *The Constitution of the Roman Republic* (Oxford: Oxford University Press).

Linz, Juan J. (1998), 'Democracy's Time Constraints', *International Political Science Review* 19 (1), 19–37.

Lokdam, Hjalte (in preparation), *Banking on Sovereignty – The Political Theory of Central Bank Independence and the European Central Bank*, PhD thesis (London: LSE).

Lorimer, Marta (in preparation), *Europe from the Far Right: Europe in the Ideology of the Front National and Movimento Sociale Italiano / Alleanza Nazionale (1978–2017)*, PhD thesis (London: LSE).

MacCormick, Neil (1999), *Questioning Sovereignty: Law, State, and Nation in the European Commonwealth* (Oxford, Oxford University Press).

Machiavelli, Niccolò (2008/1531), *Discourses on Livy*, trans. Julia Conaway Bondanella and Peter Bondanella (Oxford: Oxford University Press).

Mckenzie, Lisa (2017), 'The Class Politics of Prejudice: Brexit and the Land of No-hope and Glory', *British Journal of Sociology* 68 (S1), S265–S280.

Mair, Peter (2007), 'Political Opposition and the European Union', *Government & Opposition* 41 (1), 1–17.

Mair, Peter (2013), *Ruling the Void: The Hollowing of Western Democracy* (London: Verso).

Majone, Giandomenico (1996), 'Temporal Consistency and Policy Credibility: Why Democracies Need Non-Majoritarian Institutions', *European University Institute, Working Paper* RSC No 96/57.

Mandel, Ernest (1974), 'EEC: The Common Market in Crisis', *Inprecor* 2 (June 20), 14–18.
Manin, Bernard (2008), 'The Emergency Paradigm and the New Terrorism: What If the End of Terrorism Was Not in Sight?', in Sandrine Baume and Biancamaria Fontana (eds.), *Les usages de la séparation des pouvoirs* (Paris: Michel Houdiard), 136–71.
Marcussen, Martin (2009), 'Scientization of Central Banking: The Politics of A-Politicization,' in Kenneth Dyson and Martin Marcussen (eds), *Central Banks in the Age of Euro: Europeanization, Convergence and Power* (Oxford: Oxford University Press), 373–90.
Marshall, Geoffrey (1987), *Constitutional Conventions: The Rules and Forms of Political Accountability* (Oxford: Oxford University Press).
Matthijs, Matthias (2016), 'Powerful Rules Governing the Euro: The Perverse Logic of German Ideas', *Journal of European Public Policy*, 23 (3), 375–91.
Menéndez, Agustín José (2013), 'The Existential Crisis of the European Union', *German Law Journal* 14 (5), 453–526.
Menéndez, Agustín José (2016), 'The Refugee Crisis: Between Human Tragedy and Symptom of the Structural Crisis of European Integration', *European Law Journal* 22 (4), 388–416.
Menéndez, Agustín José (2017), 'The Crisis of Law and the European Crises: From the Social and Democratic *Rechtsstaat* to the Consolidating State of (Pseudo-)Technocratic Governance', *Journal of Law and Society* 44 (1), 56–78.
Mills, C. Wright (1956/2000), *The Power Elite* (Oxford: Oxford University Press), Chap. 1, 'The Higher Circles'.
Milstein, Brian (2015), 'Thinking Politically about Crisis: A Pragmatist Perspective', *European Journal of Political Theory* 14 (2), 141–60.
Milward, Alan (1992), *The European Rescue of the Nation-State* (London: Routledge).
Mischner, Sabine (2018), 'The Temporalities of Exception. The Long Shadow of the American Civil War', *Zeitschrift für Politikwissenschaft* 28 (4), 439–451.
Mitrany, David (1941/1975), 'A War-Time Submission', in *The Functional Theory of Politics* (London: Martin Robertson).
Mody, Ashoka (2018), *EuroTragedy: A Drama in Nine Acts* (Oxford: Oxford University Press).
Moffitt, Benjamin (2016), *The Global Rise of Populism: Performance, Political Style, and Representation* (Stanford: Stanford University Press).
Monnet, Jean (1978), *Memoirs*, trans. Richard Mayne (London: Collins).
Moravcsik, Andrew (1998), *The Choice for Europe: Social Purpose and State Power from Messina to Maastricht* (Ithaca: Cornell University Press).
Mouffe, Chantal (2005), *On the Political* (London: Routledge).
Mourlon-Druol, Emmanuel (2010), 'Filling the EEC Leadership Vacuum? The Creation of the European Council in 1974', *Cold War History* 10 (3), 315–39.
Mourlon-Druol, Emmanuel (2010), 'The Victory of the Intergovernmental Method? The Emergence of the European Council in the Community's Institutional Set-up (1974–77)', in Daniela Preda, Daniele Pasquinucci (eds), *The Road Europe Travelled Along: The Evolution of the EEC/EU Institutions and Policies* (Brussels: Peter Lang), 17–40.
Mourlon-Druol, Emmanuel (2012). 'Managing from the Top: Globalisation and the Rise of Regular Summitry, Mid-1970s–Early 1980s', *Diplomacy and Statecraft* 23 (4), 679–703.
Mourlon-Druol, Emmanuel (2016), 'Steering Europe: Explaining the Rise of the European Council, 1975–1986', *Contemporary European History* 25 (3), 409–37.
Moury, Catherine and Adam Standring (2017), '"Going beyond the Troika": Power and Discourse in Portuguese Austerity Politics', *European Journal of Political Research* 56 (3), 660–79.

Mudde, Cas (2004), 'The Populist Zeitgeist,' *Government and Opposition*, 39 (3), 541–63.

Mudde, Cas and Cristobal Rovira Kaltwasser (2017), *Populism: A Very Short Introduction* (Oxford: Oxford University Press).

Mudge, Stephanie (2018), *Leftism Reinvented: Western Parties from Socialism to Neoliberalism* (Cambridge MA: Harvard University Press).

Mudge, Stephanie and Antoine Vauchez (2012), 'Building Europe on a Weak Field: Law, Economics, and Scholarly Avatars in Transnational Politics', *American Journal of Sociology* 118 (2), 449–92.

Mudge, Stephanie and Antoine Vauchez (2016), 'Fielding Supranational: The ECB Hyper-scientization as a Field Effect', *Sociological Review* 64 (2).

Müller, Jan-Werner (2016), *What is Populism?* (Philadelphia: University of Pennsylvania Press).

NEON, NEF, FrameWorks Institute, PIRC (2018), Framing the Economy: How to Win the Case for a Better System (London).

Niemann, Arne and Natascha Zaun (2018), 'EU Refugee Policies and Politics in Times of Crisis: Theoretical and Empirical Perspectives', *Journal of Common Market Studies* 56 (1), 3–22.

Niemann, Arne and Johanna Speyer (2018), 'A Neofunctionalist Perspective on the "European Refugee Crisis": The Case of the European Border and Coast Guard', *Journal of Common Market Studies* 56 (1), 23–43.

Niesen, Peter (2019), 'Reframing Civil Disobedience: Constituent Power as a Language of Transnational Protest', *Journal of International Political Theory* 15 (1), 31–48.

Nippel, Wilfried (2012), 'Saving the Constitution: The European Discourse on Dictatorship', in Janet Coleman and Paschalis M. Kitromilides (eds), *In the Footsteps of Herodotus: Towards European Political Thought* (Florence: Leo S. Olschki), 29–50.

Nisbet, Robert (1994), *History of the Idea of Progress* (London: Routledge).

Noelle-Neumann, Elizabeth (1984), *The Spiral of Silence: Public Opinion–Our Social Skin* (Chicago: University of Chicago Press).

Ochoa Espejo, Paulina (2017), 'Populism and the Idea of the People', in Cristobal Rovira Kaltwasser, Paul Taggart, Paulina Ochoa Espejo, and Pierre Ostiguy (eds), *The Oxford Handbook of Populism* (Oxford: Oxford University Press), 607–28.

Offe, Claus (2009), 'Governance: An "Empty Signifier"?', *Constellations* 16 (4), 550–62.

Offe, Claus (2014), *Europe Entrapped* (Cambridge: Polity).

Offe, Claus (2015), 'Temporalstrukturen sozialer Macht', *Leviathan: Berliner Zeitschrift für Sozialwissenschaft* 43 (Sonderband 30), 29–51.

Olson, Richard G. (2015), *Scientism and Technocracy in the Twentieth Century: The Legacy of Scientific Management* (Lanham: Rowman & Littlefield).

Olsson, E.-K. and K. Hammargård (2016), 'The Rhetoric of the President of the European Commission: Charismatic Leader or Neutral Mediator?' *Journal of European Public Policy* 23 (4), 550–70.

Opitz, Sven and Ute Tellmann (2015), 'Future Emergencies: Temporal Politics in Law and Economy', *Theory, Culture and Society* 32 (2), 107–29.

Orr, Graeme (2015), *Ritual and Rhythm in Electoral Systems: A Comparative Legal Account* (Farnham: Ashgate).

Pansardi, Pamela and Francesco Battegazzorre (2018), 'The Discursive Legitimation Strategies of the President of the Commission: A Qualitative Content Analysis of the State of the Union Addresses (SOTEU)', *Journal of European Integration*, 40 (7), 853–71.

Pansardi, Pamela and Pier Domenico Tortola (2019), 'The Charismatic Leadership of the ECB Presidency: A Language-based Analysis', *European Journal of Political Research* 58 (1), 96–116.

Patberg, Markus (2019), 'Destituent Power in the European Union: On the Limits of a Negativistic Logic of Constitutional Politics', *Journal of International Political Theory* 15 (1), 82–99.

Patberg, Markus (in preparation), '*Constituent Power in the European Union* (Oxford: Oxford University Press).

Paudyn, Bartholomew (2013), 'Credit Rating Agencies and the Sovereign Debt Crisis: Performing the Politics of Creditworthiness through Risk and Uncertainty', *Review of International Political Economy* 20 (4), 788–818.

Picon, Antoine (2007), 'French Engineers and Social Thought, 18–20th Centuries: An Archeology of Technocratic Ideals', *History and Technology* 23 (3), 197–208.

Posner, Eric A. and Adrian Vermeule (2011), *The Executive Unbound: After the Madisonian Republic* (Oxford: Oxford University Press).

Pound, Roscoe (1913), 'Justice According to Law', *Columbia Law Review*, 13 (8), 696–713.

Preuß, Ulrich K. (2017), 'Die Krise der Europäischen Union als Ausnahmezustand?', *KJ Kritische Justiz* 50, 51–67.

Putnam, Robert (1977), 'Elite Transformation in Advanced Industrial Societies: An Empirical Assessment of the Theory of Technocracy', *Comparative Political Studies* 10 (3), 383–412.

Radaelli, Claudio (1999), *Technocracy in the European Union* (London: Routledge).

Raithby, John (ed.) (1819), *Statutes of the Realm: Volume 5, 1628–80* (British History online).

Ramraj Victor V. (ed.) (2008), *Emergencies and the Limits of Legality* (Cambridge: Cambridge University Press).

Rawls, John (1999 [1971]), *A Theory of Justice* (Cambridge MA: Harvard University Press).

Reckwitz, Andreas (2017), *Die Gesellschaft der Singularitäten–Zum Strukturwandel der Moderne* (Berlin: Suhrkamp).

Rehn, Olli and Jörg Asmussen (2012), 'Responses to Written Questions on the Economic and Social Crisis in Greece by Several Political Groups within the European Parliament', 23 March. https://www.ecb.europa.eu/press/key/date/2012/html/sp120327.pdf?1e3c8a4aca0d753e511f7cc31170a232.

Reinhart, Carmen and Kenneth Rogoff (2009), *This Time is Different: Eight Centuries of Financial Folly* (Princeton: Princeton University Press).

Rhinard, Mark (2019), 'The Crisification of the EU', *Journal of Common Market Studies* 57 (3), 616–633.

Riescher, Gisela (1994), *Zeit und Politik: Zur institutionellen Bedeutung von Zeitstrukturen in parlamentarischen und präsidentiellen Regierungssystemen* (Baden-Baden: Nomos).

Riker, William (1986), *The Art of Political Manipulation* (New Haven: Yale University Press).

Rittberger, Berthold (2014), 'Integration without Representation? The European Parliament and the Reform of Economic Governance in the EU', *Journal of Common Market Studies* 52 (6), 1174–83.

Roberts, Kenneth M. (2017), 'Populism and Political Parties', in Rovira Kaltwasser, Taggart, Ochoa Espejo, and Ostiguy (eds), *The Oxford Handbook of Populism* (Oxford: Oxford University Press).

Rödl, Florian (2012), 'EU im Notstandsmodus', *Blätter für Deutsche und Internationale Politik* 5 (5) (https://www.blaetter.de/archiv/jahrgaenge/2012/mai/eu-im-notstandsmodus).

Rosa, Hartmut (2013), *Social Acceleration: A New Theory of Modernity* (New York: Columbia University Press).

Rosa, Hartmut and William Scheuerman (eds) (2009), *High-Speed Society* (University Park: Pennsylvania State University Press).

Rosenboim, Or (2017), *The Emergence of Globalism: Visions of World Order in Britain and the United States, 1939–1950* (Princeton: Princeton University Press).

Rossiter, Clinton (1948), *Constitutional Dictatorship: Crisis Government in the Modern Democracies* (Princeton: Princeton University Press).

Rovira Kaltwasser, Cristobal, Paul Taggart, Paulina Ochoa Espejo, and Pierre Ostiguy (2017), 'Populism: An Overview of the Concept and the State of the Art', in Rovira Kaltwasser, Taggart, Ochoa Espejo, and Ostiguy (eds), *The Oxford Handbook of Populism* (Oxford: Oxford University Press), 1–26.

Rüb, Friedbert W. (2015), 'Die Zeit und die Unzeit der Politik', *Leviathan: Berliner Zeitschrift für Sozialwissenschaft* 43 (Sonderband 30), 195–217.

Rubenstein, Jennifer (2015), 'Emergency Claims and Democratic Action', *Social Philosophy and Politics* 32 (1), 101–26.

Ruiz-Rufino, Rubén and Sonia Alonso (2017), 'Democracy without Choice: Citizens' Perceptions of Government Autonomy during the Eurozone Crisis', *European Journal of Political Research* 56, 320–45.

Rummens, Stefan (2017), 'Populism as a Threat to Liberal Democracy', in Rovira Kaltwasser, Taggart, Ochoa Espejo, and Ostiguy (eds), *The Oxford Handbook of Populism* (Oxford: Oxford University Press), 554–70.

Ryle, Gilbert (1949), 'Knowing How and Knowing That', in *Gilbert Ryle: Collected Papers, vol. 2.* (New York: Barnes and Noble).

Sanchez-Cuenca, Ignacio (2017), 'From a Deficit of Democracy to a Technocratic Order: The Postcrisis Debate on Europe', *Annual Review of Political Science* 20, 351–69.

Sandbu, Martin (2015), *Europe's Orphan: The Future of the Euro and the Politics of Debt* (Princeton: Princeton University Press).

Schäfer, David (2016), 'A Banking Union of Ideas? The Impact of Ordoliberalism and the Vicious Circle on the EU Banking Union', *Journal of Common Market Studies* 54 (4), 961–80.

Scharpf, Fritz (1999), *Governing in Europe: Effective and Democratic?* (Oxford: Oxford University Press).

Scharpf, Fritz (2006), 'The Joint Decision Trap Revisited', *Journal of Common Market Studies* 44 (4), 845–64.

Scharpf, Fritz (2014), 'Political Legitimacy in a Non-optimal Currency Area', in Sara Hobolt and Olaf Cramme (eds), *Democratic Politics in a European Union under Stress* (Oxford: Oxford University Press), 19–47.

Schäuble, Wolfgang (2011), 'We Can Only Achieve a Political Union If We Have a Crisis', *New York Times*, 18 November (http://www.nytimes.com/2011/11/19/world/europe/for-wolfgang-schauble-seeing-opportunity-in-europes-crisis.html).

Schedler, Andreas and Javier Santiso (1998), 'Democracy and Time: An Invitation', *International Political Science Review* 19 (1), 5–18.

Schelkle, Waltraud (2017), *The Political Economy of Monetary Stability: Understanding the Euro Experiment* (Oxford: Oxford University Press).

Scheppele, Kim Lane (2010), 'Exceptions that Prove the Rule: Embedding Emergency Governing in Everyday Constitutional Life', in Jeffrey K. Tulis and Stephen Macedo (eds), *The Limits of Constitutional Democracy* (Princeton: Princeton University Press), 124–54.

Scheppele, Kim Lane (2012), 'The New Judicial Deference', *Boston University Law Review* 92, 89–170.

Scheppele, Kim Lane (2018), 'Autocratic Legalism', *University of Chicago Law Review* 85 (2), 545–83.

Scheuerman, William (1999), 'The Economic State of Emergency,' *Cardozo Law Review* 21 (5–6), 1869–94.

Scheuerman, William (2004), *Liberal Democracy and the Social Acceleration of Time* (Baltimore: Johns Hopkins University Press).

Scheuerman, William (2006), 'Survey Article: Emergency Powers and the Rule of Law After 9/11', *Journal of Political Philosophy* 14 (1), 61–84.

Scheuerman, William (2015), 'Recent Theories of Civil Disobedience: An Anti-Legal Turn?', *Journal of Political Philosophy* 23 (4), 427–49.

Scheuerman, William (2016), 'Crises and Extra-Legality: From Above and from Below', in Poul F. Kjaer and Niklas Olsen (eds), *Critical Theories of Crisis in Europe: From Weimar to the Euro* (Lanham MD: Rowman & Littlefield), 197–212.

Scheuerman, William (2017), '*Die Globalisierung von Carl Schmitt?*', *KJ Kritische Justiz* 50, 30–7.

Scheuerman, William (2017), 'States of Emergency', in Jens Meierhenrich and Oliver Simons (eds), *Oxford Handbook of Carl Schmitt* (Oxford: Oxford University Press), 547–69.

Scheuerman, William (2018), *Key Concepts: Civil Disobedience* (Cambridge: Polity).

Scheuerman, William (2019), 'Constituent Power and Civil Disobedience: Beyond the Nation State?', *Journal of International Political Theory* 15 (1), 49–66.

Schimmelfennig, Frank (2018), 'European Integration (Theory) in Times of Crisis: A Comparison of the Euro and Schengen Crises', *Journal of European Public Policy* 25 (7), 969–89.

Schmidt, Vivien (2018), 'Rethinking EU Governance: From "Old" to "New" Approaches to Who Steers Integration', *Journal of Common Market Studies* 56 (7), 1544–1561.

Schmitt, Carl (1921/2013), *Dictatorship* (Cambridge: Polity).

Schmitt, Carl (1928/2008), *Constitutional Theory* (Durham NC: Duke University Press).

Schmitt, Carl (2005), *Political Theology: Four Chapters on the Concept of Sovereignty* (Chicago: Chicago University Press).

Schmitter, Philippe C. (1971), 'A Revised Theory of Regional Integration', in Leon N. Lindberg and Stuart A. Scheingold (eds), *Regional Integration: Theory and Research* (Cambridge, MA: Harvard University Press), 232–64.

Schmitter, Philippe C. (2004), 'Neo-neo-functionalism?', in Antje Wiener and Thomas Diez (eds), *European Integration Theory* (Oxford: Oxford University Press), 45–74.

Schrag Sternberg, Claudia (2013), *The Struggle for EU Legitimacy, 1950–2005* (Basingstoke: Palgrave Macmillan).

Scicluna, Nicole (2014), 'Politicization Without Democratization: How the Eurozone Crisis is Transforming EU Law and Politics', *International Journal of Constitutional Law* 12 (3), 545–71.

Scicluna, Nicole (2015), *European Union Constitutionalism in Crisis* (London: Routledge).

Scicluna, Nicole (2018), 'Integration through the Disintegration of Law? The ECB and EU Constitutionalism in the Crisis', *Journal of European Public Policy* 25 (12), 1874–1891.

Scipioni, Marco (2018), '*De Novo* Bodies and EU Integration: What is the Story behind EU Agencies' Expansion?', *Journal of Common Market Studies* 56 (4), 768–84.

Scipioni, Marco (2018), 'Failing Forward in EU Migration Policy? EU Integration after the 2015 Asylum and Migration Crisis', *Journal of European Public Policy* 28 (9), 1357–1375.

Seabrooke, Leonard (2014), 'Identity Switching and Transnational Professionals', *International Political Sociology* 8 (3), 335–7.

Seikel, Daniel (2016), 'Flexible Austerity and Supranational Autonomy: The Reformed Excessive Deficit Procedure and the Asymmetry between Liberalization and Social Regulation in the EU', *Journal of Common Market Studies* 54 (6), 1398–416.

Sending, Ole Jacob (2014), 'The International Civil Servant', *International Political Sociology* 8 (3), 338–40.
Seville, Astrid (2017), *"There is No Alternative": Politik zwischen Demokratie und Sachzwang* (Frankfurt am Main: Campus).
Skinner, Quentin (1998), *Liberty before Liberalism* (Cambridge: Cambridge University Press).
Slobodian, Quinn (2018), *Globalists: The End of Empire and the Birth of Neoliberalism* (Cambridge MA: Harvard University Press).
Smith, Alastair (2004), *Electoral Timing* (Cambridge: Cambridge University Press).
Snyder, Timothy (2010), *Bloodlands: Europe Between Hitler and Stalin* (New York: Basic Books).
Sorell, Tom (2013), *Emergencies and Politics: A Sober Hobbesian Approach* (Cambridge: Cambridge University Press).
Spinelli, Altiero and Ernesto Rossi (1941), 'The Manifesto of Ventotene for a Free and United Europe' (https://www.cvce.eu/content/publication/1997/10/13/316aa96c-e7ff-4b9e-b43a-958e96afbecc/publishable_en.pdf
Stavrakakis, Yannis, Giorgos Katsambekis, Alexandros Kioupkiolis, Nikos Nikisianis and Thomas Siomos (2018), 'Populism, Anti-populism and Crisis', *Contemporary Political Theory* 17 (1), 4–27.
Storey, Andy (2019), 'Authoritarian Neoliberalism in Europe: The Red Herring of Ordoliberalism', *Critical Sociology* (online early).
Streeck, Wolfgang (2013), *Gekaufte Zeit. Die vertagte Krise des demokratischen Kapitalismus* (Berlin: Suhrkamp).
Streeck, Wolfgang (2014), 'The Politics of Public Debt: Neoliberalism, Capitalist Development and the Restructuring of the State', *German Economic Review* 15 (1), 143–65.
Streeck, Wolfgang (2016), *How Will Capitalism End?* (London: Verso).
Sunstein, Cass (1996), *Legal Reasoning and Political Conflict* (Oxford: Oxford University Press).
Suntrup, Jan (2018), 'From Emergency Politics to Authoritarian Constitutionalism? The Legal and Political Costs of EU Financial Crisis Management', *German Law Journal* 19 (2), 375–402.
Tamanaha, Brian Z. (2004), *On the Rule of Law: History, Politics, Theory* (Cambridge: Cambridge University Press).
Teitel, Ruti (1997), 'Transitional Jurisprudence: The Role of Law in Political Transformation', *Yale Law Journal* 106 (7), 2009–80.
Thompson, Dennis (2004), 'Election Time: Normative Implications of Temporal Properties of the Electoral Process in the United States', *American Political Science Review* 98 (1), 51–63.
Tooze, Adam (2018), *Crashed: How a Decade of Financial Crises Changed the World* (London: Penguin).
Tsipras, Alexis (2015) 1 July 2015 (https://archive.change4all.eu/change-in-words/detail/prime-minister-alexis-tsipras-message.html/).
Tucker, Paul (2018), *Unelected Power: The Quest for Legitimacy in Central Banking and the Regulatory State* (Princeton: Princeton University Press).
Tuori, Kaarlo and Klaus Tuori (2014), *The Eurozone Crisis: A Constitutional Analysis* (Cambridge: Cambridge University Press).
Turner, Stephen (2001), 'What Is the Problem with Experts?', *Social Studies of Science* 31 (1), 123–49.
Turner, Stephen (2014), *The Politics of Expertise* (London: Routledge).

Tushnet, Mark (2005), 'Emergencies and the Idea of Constitutionalism', in Tushnet (ed.), *The Constitution in Wartime: Beyond Alarmism and Complacency* (Durham NC: Duke University Press), 39–54.

UNHCR (2015), 'Over One Million Sea Arrivals Reach Europe in 2015', 30 Dec. (http://www.unhcr.org/afr/news/latest/2015/12/5683d0b56/million-sea-arrivals-reach-europe-2015.html).

Van Middelaar, Luuk (2019), *Alarums and Excursions: Improvising Politics on the European Stage* (Newcastle: Agenda).

Van Rompuy, Herman, with José Manuel Barroso, Jean-Claude Juncker, and Mario Draghi (2012), 'Towards a Genuine Economic and Monetary Union' ['Four Presidents' Report'] (Brussels: Council).

Van 't Klooster, Jens (2018), 'Democracy and the European Central Bank's Emergency Powers', *Midwest Studies in Philosophy* 42 (1), 270–91.

Vanberg, Viktor J. (2004), 'The Freiburg School: Walter Eucken and Ordoliberalism', Paper 04/11 (Freiburg: Walter Eucken Institut).

Varoufakis, Yanis (2016), 'Europe's Left after Brexit', 8 September (https://yanisvaroufakis.eu/2016/09/08/europes-left-after-brexit/).

Varoufakis, Yanis (2017), *Adults in the Room: My Battle with Europe's Deep Establishment* (London: Penguin).

Vauchez, Antoine (2008), 'The Force of a Weak Field: Law and Lawyers in the Government of the European Union', *International Political Sociology* 2 (2), 128–44.

Vauchez, Antoine (2015), *Brokering Europe. Euro-lawyers and the Making of a Transnational Polity* (Cambridge: Cambridge University Press).

Vauchez, Antoine (2016), *Democratizing Europe* (Basingstoke: Palgrave Macmillan).

Vogl, Joseph (2016), *The Ascendancy of Finance* (Cambridge: Polity).

Von Bogdandy, Armin (2018), 'Ways to Frame the European Rule of Law: *Rechtsgemeinschaft*, Trust, Revolution, and Kantian Peace', *European Constitutional Law Review* 14, 675–99.

Von Bogdandy, Armin and Michael Ioannidis (2014), 'Systemic Deficiency in the Rule of Law: What It Is, What Has Been Done, What Can Be Done', *Common Market Law Review* 51, 59–96.

Wagner, Rikke (2015), 'Transnational Civil Dis/obedience in the Danish Family Unification Dispute', *European Political Science Review* 7 (1), 43–62.

Watson, Matthew and Colin Hay (2004), 'The Discourse of Globalization and the Logic of No Alternative: Rendering the Contingent Necessary in the Political Economy of New Labour', *Policy and Politics* 30 (4), 289–305.

Weber, Max (1968), 'The Nature of Charismatic Authority and its Routinization', in S. N. Eisenstadt (ed.) *Max Weber on Charisma and Institution Building: Selected Papers* (Chicago: University of Chicago Press).

Weiler, J. H. H. (2012), 'Integration through Fear', *European Journal of International Law* 23 (1), 1–5.

Wessels, Wolfgang (2015), *The European Council* (Basingstoke: Palgrave Macmillan).

White, Jonathan (2010), 'Europe and the Common', *Political Studies* 58 (1), 104–22.

White, Jonathan (2010), 'The Politics of Other Citizens', *Citizenship Studies* 14 (4), 411–27.

White, Jonathan (2010), 'Responding to Norm Indeterminacy Outside the Nation-State Frame', *Comparative Sociology* 9 (5), 611–30.

White, Jonathan (2011), *Political Allegiance after European Integration* (Basingstoke: Palgrave Macmillan).

White, Jonathan (2013), 'Left and Right in the Economic Crisis', *Journal of Political Ideologies* 18 (2), 150–70.
White, Jonathan (2014), 'Politicizing Europe: The Challenge of Executive Discretion', in Sara Hobolt and Olaf Cramme (eds), *Democratic Politics in a European Union under Stress* (Oxford: Oxford University Press), 87–102.
White, Jonathan (2015), 'Emergency Europe', *Political Studies* 63 (2), 300–18.
White, Jonathan (2017), 'Between Rules and Discretion: Thoughts on Ordo-Liberalism', in Josef Hien and Christian Joerges (eds), *Ordo-Liberalism, Law and the Rule of Economics* (Oxford: Hart), 289–300.
White, Jonathan (2017), 'Revisionism as a Logic of Institutional Change', *European Law Journal* 23 (5), 406–16.
White, Jonathan and Lea Ypi (2016), *The Meaning of Partisanship* (Oxford: Oxford University Press).
Wolkenstein, Fabio (2016), 'A Deliberative Model of Intra-Party Democracy', *Journal of Political Philosophy* 24 (3), 297–320.
Wolkenstein, Fabio (2018), 'Democracy, Transnational Partisanship and the EU', *Journal of Common Market Studies* 56 (2), 284–99.
Wolkenstein, Fabio (2018), 'Transnational Partisanship and Networked Constituent Power in the EU', *LEQS Paper No. 138/2018.*
Woodruff, David M. (2016), 'Governing by Panic: The Politics of the Eurozone Crisis', *Politics & Society* 44 (1), 81–116.
Zerubavel, Evitar (1985), *Hidden Rhythms: Schedules and Calendars in Social Life* (Berkeley: University of California Press).
Zhukov, Yuri (2016), 'NATO's Mediterranean Mission: What the Alliance Is Doing in the Aegean Sea', *Foreign Affairs* 21 February.

Index

Note: Tables are indicated by an italic '*t*' following the page number, end of chapter notes are indicated by a letter 'n' between page number and note number.

For the benefit of digital users, indexed terms that span two pages (e.g., 52–53) may, on occasion, appear on only one of those pages.

acceleration *see* social acceleration
agency *see* political agency
Alternative für Deutschland 138
American Federalists 55–6, 57n.51, 167–9
anterior normal 24–6, 86, 132–3
anti-populism 142–6
Arab Socialist Baath Party 184
asylum system 81, 83–4, 160–1
austerity measures 27–8, 50–1, 94, 97–8
authority 5–6, 10, 62, 66, 77, 85, 103–4, 159, 192, 203–4
  charismatic 122–4
  democratic 4
  of market agents 69–70
  procedural 11, 86, 98–104, 150
  and securitization 66, 69–73, 76–7
  technocratic 11–12, 86, 106–26
authoritarian drift 102–3
authoritarian rule 3, 62–3, 86, 142, 176

Barroso, José Manuel 22–3
Blair, Tony 131–2
Bretton Woods system 57, 69–70
Brexit 7, 15, 135, 137–8, 150, 197–8
Britain *see* United Kingdom
Brown, Wendy 131–2

capitalism, financial 50, 67–8
'cartel parties' 174–5
casuistry 118
Catholic Church 51–2
charisma 122–3, 122n.56, 124
Chicago School economics 64–5, 75, 131–2
Christian Democracy 45–6, 51–2
civil disobedience 155, 158–9, 161, 177n.26,
  see also principled disobedience
CJEU *see* European Court of Justice (CJEU)
Coal and Steel Community 40, 43
coercion 87, 101–3
commissarial dictatorship 198–9
common-market programme 41, 74, 78–9
  and creation of European Council 56–60
  'four freedoms' 41, 43, 80
  primacy of policy ends 41, 46–51
Community Method 47–8, 58–9, 61, 133–4
conditional loans 20
consequentialist forms of authority 103–4
constitutional conventions 19
constitutional order 89, 164–5
  commitment to 161–2, 165–6
constitutional reassurance
  paths to 87, 90–4
  problem of 88–90
contestation of emergency rule 147–62, 165–6
  judicial responses 151–2
  legislative responses 150–1
  malicious compliance 153
  principled disobedience 147–9, 154–62, 164–6, 172
  reflection on legitimacy of emergency rule 162–5
  see also political constraints on emergency rule
Copenhagen School 71–2
Court of Auditors *see* European Court of Auditors
Court of Justice *see* European Court of Justice (CJEU)
crisis expertise 114–16, 123–5, 188
crisis fatigue 101

deadlines 54, 60, 199
debt 68–71
  see also Euro crisis
debt restructuring 50–1
decision-making
  acceleration of 6, 21, 23–4, 47n.21, 110–12
  and deadlines 54, 60, 199
  and diffusion of power 51–6
  and technocracy 110–12
de-institutionalization 26, 29, 38, 112, 133–4, 168

democracy 3, 8, 47–8
 challenging norms of 21–2
 intra-party 171–2
 populist appeal to 137–9
 and the unique 189–95
depoliticization 23, 44, 77–8, 145, 162–3, 174–5, 199–200
Dewey, John 178–9
dictatorship 30–2, 55–6, 80n.48, 100, 191, 198–9
DiEM25 movement 156, 158, 183–6
diffusion of power 41, 51–6, 57n.51, 78–9, 167–9
 and Euro crisis 26–9, 34–5
disaster capitalism 64
disasters 64, 67, 72–3, 91
disavowal of agency 6–7, 127–8, 130–4
discourse of exceptionalism 18–26, 189–95
discretion, executive 5, 14, 38–9, 46–56, 74–5, 88–94, 114–15, 118, 123–4, 130, 139–40, 151, 164–5, 171–2, 194–5
disobedience 147–9, 154–62, 164–6, 172
 normative foundations 158–62
Draghi, Mario 51n.33, 115, 124n.61, 150n.9
dual state concept 100–1
Dublin system 81, 83–4
Duchêne, Francois 40–1

ECB *see* European Central Bank (ECB)
ECI *see* European Citizens' Initiative (ECI)
economic expertise 75, 77, 123–4
economic recession 57
emergency rule 2–8
 appeal to exceptional circumstances 22–6, 30, 33–4, 37–9, 109–10, 162–5, 188–204
 appeal to necessity 6–7, 22–3, 26, 61–2, 82, 128–35, 154–5, 160, 162–3
 challenge to technocracy 107–13
 claims about conclusion of 87, 90–8
 coercive legacy of 102–3
 conceptual foundations 30–5
 constitutional reassurance problem 88–90
 informal 31–2
 judicial responses to 151–2
 know-that and know-how 120–2
 legislative responses to 150–1
 legitimacy of 3, 162–5
 malicious compliance 153
 as opportunity for expertise 113–17
 populism as emergency-to-come 142–5
 procedural authority problems 98–101
 and public scrutiny 61–2
 Roman model 30–1, 33–4, 80n.48, 89, 191
 see also contestation of emergency rule; EU emergency politics; political constraints on emergency rule
Empty Chair Crisis 57
engineers 113–14, 136n.26
English Triennial Act (1641) 177
Enroth, Henrik 74–5
epistemic authority 152
equality of states 20
EU emergency politics 64–85, 199–204
 beyond the Euro Crisis 78–85
 claims about conclusion of 94–8
 coercive legacy of 102
 constitutional reassurance problem 89–90
 governance paradigm 73–9, 83–4, 164, 170–1, 174–5
 legitimacy of 163–5
 'migration crisis' 80–5
 procedural authority problems 98–101
 rise of extreme conditions 67–73
 and search for non-procedural authority 102–4
 see also Euro crisis; political constraints on emergency rule
Euro crisis 5–6, 16–39, 50–1, 64–5, 71, 79–80, 84, 179
 and appeals to necessity 22–3, 26, 129
 and concept of emergency rule 30–5
 escaping norms 18–22, 30–2
 exceptionalism 22–6, 33–4, 193, 196–8
 executive activism 26–30, 34–5
 and normality-to-come 94–6
 and technocracy 110–12, 114–16, 120–1, 123–4
 and theories of emergency rule 35–9
Eurogroup 18–19, 23, 27–8, 196–7
European Border and Coast Guard (EBCG) 82–3
European Central Bank (ECB) 20, 23, 38, 46, 50–1
 crisis expertise 115, 123–4
 neoliberalism 64–5
 and normality-to-come 96–7, 103–4
 technocracy 110–11, 114–15, 119–20, 123–4, 124n.59
 see also Troika
European Citizens' Initiative (ECI) 181–2, 185
European Commission 28, 43, 57, 61, 82–3, 150
 crisis expertise 114–15, 125
 economic expertise 75
 neoliberalism 64–5
 technocracy 110–11, 114–15, 119–20
 see also Troika
European Council 8, 61, 75, 125, 150, 196
 creation of 42, 56–60
 and partisanship 171–2, 183
 and the Troika 17, 28
European Court of Auditors 28, 110–11
European Court of Justice (CJEU) 48, 96, 114–15, 123–4, 124n.59, 151–2

European integration 4–5, 15, 40–63, 74, 78–9
and creation of European Council 56–60
diffusion of power 41, 51–6, 78–9
instrumentalist orientation 41–6
and neoliberalism 64–5
and Ordoliberalism 44–6, 48–9, 52, 131
primacy of policy ends 41, 46–51
and state sovereignty 15, 20, 51–6, 76n.38
theories of 5, 27, 29, 38–9, 42, 50
European Parliament 21, 43, 97, 140–1, 150, 172–3, 180–1, 185–6, 193
European Semester 96, 102, 143–4, 156
European Stability Mechanism (ESM) 17, 99–100, 151, 192–3
EUROPOL 82
exceptionalism 30, 37–9, 162–5, 188–204
claims about conclusion of 87, 90–8
constitutional reassurance problem 88–90, 100–1
Euro crisis 22–6, 33–4, 193, 196–8
'migration crisis' 80–5
rise of extreme conditions 67–73
and technocracy 109–10
executive activism 6
Euro crisis 26–30, 34–5
executive discretion 5, 14, 38–9, 46–56, 74–5, 88–94, 114–15, 118, 123–4, 130, 139–40, 151, 164–5, 171–2, 194–5
executive power
claim of new constraints on 92–3, 95–7
and governance paradigm 73–9, 83–4, 164, 170–1, 174–5
individualization of 28, 167–9
see also diffusion of power
'executive unbound' concept 8
expertise
crisis 114–16, 123–5, 188
economic 75, 77, 123–4
emergency rule as opportunity for 113–17
judicial response 123–4, 152
legal 74–7
meta-expertise 123–4
see also technocracy

Far Right parties 83, 149, 160–2, 185–6
'fast policy' 68–9
Federalists, American 55–6, 57n.51, 167–9
federation 52–3, 76n.38
Fidesz party, Hungary 160–1
Fiscal Compact 5–6, 149, 153
'Five Presidents' 1, 29–30, 103–4
forecasting *see* prediction
'four freedoms' of the common market 41, 43, 80
'Four Presidents' 29–30
Fraenkel, Ernst 100
France 58, 113–14, 137–8, 193
Frankfurt group 28
Freeden, Michael 141
free movement 41, 43
and 'migration crisis' 80–5
Friedman, Milton 64
Friedrich, Carl 31
FRONTEX 82–3
Front National, France 137–8
functional constitutionalism 43–4

Germany 23–4, 33, 58, 83–4, 151
governance paradigm 73–9, 83–4, 164, 170–1, 174–5
Great Depression 106–7, 113–14, 191, 193
Greece 5–6, 20, 27, 71, 147–9, 156, 159–61
Greek referendum (2015) 159, 181
Grimm, Dieter 49–50
Guyer, Jane 200

Hallstein, Walter 40
Hamilton, Alexander 55–6, 57n.51, 80n.48, 167–9
Hay, Colin 131–2
Hayek, Friedrich von 52, 113–14
heresthetics of emergency 95
hierarchy *see* non-hierarchical constitutional structure
Honig, Bonnie 148–9
Hungary 160–1

ideology 64, 78, 171, 202–3
Christian Democracy 45–6, 51–2
neoliberalism 64–5, 75, 78, 128, 131–2
Ordoliberalism 44–6, 48–9, 52, 75, 131
and technocracy 108
International Monetary Fund (IMF) 28, 64–5, 78, 96–7, 114–15
see also Troika
individualization of power 28, 167–9
individual rights 8
institutional periodicity 175–80, 182–5
instrumentalism 41–6, 66
governance paradigm 73–9, 83–4, 164, 170–1, 174–5
see also technocracy
intergovernmentalism 5, 27, 29, 38–9, 42
intermediary agents 66, 69–73, 76–7
international law 54–5
intra-party democracy 171–2
Ipsen, Hans-Peter 43–4

Isiksel, Turkuler 43–4, 47
Italy 5–6, 20, 27, 95, 115, 121, 129

Jessop, Bob 67–8
Johnson, Boris 7, 139
judicial power, marginalization of 96–7
judicial responses to emergency rule 151–2

Keynesianism 65
Kjaer, Poul 100
Klein, Naomi 64–5, 78
knowledge
  as know-how 107, 117–22, 118*t*, 122–6
  as know-that 107, 117–22, 118*t*, 122–3, 125, 132–3
Kreuder-Sonnen, Christian 36–7

Lagarde, Christine 97, 121
law, rule of 3, 8, 47–8, 88
  commitment to 161–2
legal expertise 74–7
legislative responses to emergency rule 150–1
Le Pen, Marine 137
liberal-economic thought 44–6, 48–9, 52
Lisbon Treaty (2007/9) 59
loan defaults 70–1

Machiavelli, Niccolò 30–1, 33–4, 55, 87, 145
Madison, James 55–6
malicious compliance 153
managerialism 66, 73–9, 83–4, 170–1
market agents 66, 69–73, 76–7
Mattarella, Sergio 5–6, 95
Memoranda of Understanding (MoUs) 196–7
Merkel, Angela 23–4
meta-expertise 123–4
migration 80–5, 138, 160–1
monarchy 176–7, 180
monetary union 46
  see also Euro crisis
Monnet, Jean 3, 36, 40–1, 54, 57–9, 61
Monti, Mario 27, 115, 121, 129
Mudge, Stephanie 75

national vetoes 47–8
NATO 81–2
natural disasters 64, 67, 72–3, 91
necessity, politics of 6–7, 22–3, 26, 61–2, 82, 128–35, 154–5, 160, 162–3
neofunctionalism 5, 27, 29, 42, 50
neoliberalism 64–5, 75, 78, 128, 131–2
New Labour, UK 131–2
no-confidence votes 21, 182–3
non-hierarchical constitutional structure 41, 51–6
normality-to-come 25–6, 86–105, 132–3
normative foundations of disobedience 158–62
norms 61, 74
  escaping 18–22, 30–2, 58–9
  following versus conforming to 101
  primacy of policy ends over 41, 46–51, 75–6

Offe, Claus 176
oil shock 57
Ordoliberalism 44–6, 48–9, 52, 75, 131
over-constitutionalization 49–50

Papademos, Lucas 27, 115, 121
Paris Treaty (1951) 43
partisanship and emergency rule 170–5, 180, 183–6
periodicity of institutions 175–80, 182–5
personalization 5, 26, 29, 38, 112, 133–4, 168
policy goals, primacy of 41, 46–51, 75–6
political agency 127–46
  and appeals to necessity 6–7, 128–35
  disavowal of 6–7, 127–8, 130–4
  populism as emergency-to-come 142–5
  populism as promise of 128, 135–43, 145–6
political constraints on emergency rule 167–87
  partisanship 170–5, 180, 183–6
  and periodicity of institutions 175–80, 182–5
  temporal construction of political opposition 175–80
  threshold model 181–2, 185
  in transnational politics 180–6
political parties 66, 74–7, 202–3
  see also Far Right parties; partisanship and emergency rule
politicization 135–42, 154, 180–6
  see also depoliticization
populism 6–7, 148, 174–5, 181
  as emergency-to-come 142–5
  as promise of agency 128, 135–43, 145–6
posterior normal 25–6, 86–105, 132–3
  constitutional reassurance problem 88–90
  paths to constitutional reassurance 87, 90–4
  procedural authority problems 98–101
prediction 71, 110, 115, 117–18, 125, 177, 191
preparatory emergency management 94–6, 125
price stability goal 46, 50–1, 75, 114–15, 119–20
principled disobedience 147–9, 154–62, 164–6, 172
  normative foundations 158–62
problem-solving ethos 45, 66, 74–7, 79, 84–5
procedural authority 86–7, 98–101
procedural norms 61, 74
  escaping 18–22, 30–2, 58–9
  primacy of policy ends over 41, 46–51, 75–6
'Project Fear' 7, 135
Provisional European Government 57–60
public attitudes 15, 58–9, 61–2, 77–9, 101, 127–8, 135, 150, 154–5, 159–60, 174–5, 194, 197–8

public opinion *see* public attitudes
public scrutiny 61–2

ratings agencies 69–73, 76–7
reassurance *see* constitutional reassurance
Reckwitz, Andreas 201
referenda 159–60, 181–2
  Brexit 7, 15, 135, 137–8, 150, 197–8
refugees 80–5
revisionism 75n.32
rhetoric of exceptionalism 18–26, 189–95
rhythm, politics of 175–80, 182–5
right-wing politics 137–42, 148, 181
  see also Far Right parties
Roman Republic 30–1, 33–4, 80n.48, 89, 168, 191
Rossiter, Clinton 34, 56
rule of law 3, 8, 47–8, 88
  commitment to 161–2
Russia, troika figure 16–17
Ryle, Gilbert 117

Salvini, Matteo 140–1
Scharpf, Fritz 53–4
Schengen system 81–3
Scheuerman, William 194–5
Schmitt, Carl 31–4, 52–3, 76n.38, 100–1, 162–3
scientific method 112–14, 118*t*, 119–20
securitization studies 33–4, 71–2
securitizing agents 66, 69–73, 76–7
self-cancelling prophecy 7, 24
self-fulfilling prophecy 7
shock doctrine 64–5, 75, 78, 110, 128
silence, spiral of 154–5
Single European Act (1986/7) 59
social acceleration 67–8
sovereign debt crisis 71
sovereign dictatorship 198–9
sovereignty 5, 20
  Christian Democratic thought 51–2
  and European integration 15, 20, 51–6, 76n.38
  liberal-economic thought 52
speed 21, 23–4, 26–7, 82
spiral of silence 154–5
state sovereignty 5, 20
  Christian Democratic thought 51–2
  and European integration 15, 20, 51–6, 76n.38
  liberal-economic thought 52
subsidiarity principle 51–2
sui generis 189–90, 196–7
Syriza movement, Greece 147–9, 156, 159–61

target-setting 75–6
technocracy 86, 106–26
  defined 107–8
  emergency rule as challenge 107–13
  emergency rule as opportunity 113–17
  know-that and know-how 117–22, 118*t*
temporariness 25–6, 37, 195–6, 198–9
Thatcher, Margaret 59, 131–2
'the people' 135–7, 139, 141
There-Is-No-Alternative (TINA) doctrines 131–2
threshold model 181–2, 185
time, politics and 175–80
  deadlines 54, 60, 199
  delaying tactics 24, 179
  periodicity 175–80, 182–5
  rhythm 175–80, 182–5
  social acceleration 67–8
  sovereignty 67–8
  speed and urgency 21, 23–4, 26–7, 82
  temporariness 25–6, 37, 195–6, 198–9
  threshold model 181–2, 185
  uniqueness 24, 188–204
Tindemans, Leo 60
'too big to fail' 112
trade unions 153
transnational partisanship 173–4, 183–6
Treaty of Rome (1957) 43, 46, 58–9
Trichet, Jean-Claude 23, 33, 46n.16, 115
Troika 16–39, 64–5, 196–7
  and appeals to necessity 22–3, 26, 129
  escaping norms 18–22, 30–2
  exceptionalism 22–6
  executive activism 26–30, 34–5
  and technocracy 110–12, 114–16, 120–1, 123–4
troika imagery 16–17
Tsipras, Alexis 147–8
Turkey 81

unique circumstances 24, 188–204
  democracy and the unique 189–95,
  see also exceptionalism
United Kingdom 58, 131–2
  Brexit 7, 15, 135, 137–8, 150, 197–8
United States 33, 57, 64, 113–14, 193
  Federalists 55–6, 57n.51, 167–9

Varoufakis, Yanis 156, 196
volition *see* political agency
voluntarism 128, 135–42
votes-of-no-confidence 21, 182–3

Weber, Max 122–3
World Bank 64, 78

Zweckverband 43–4